"Maybe once every 100 years does someone emerge from the shuddering mass of humanity who speaks to us with a kind of clarity that is universally profound. Thomas Berry is such a figure. *The Great Work* will, I believe, be remembered as the touchstone, the 'bible' whose wisdom laid the groundwork for our continued healthy existence here on Earth."

Thomas Rain Crowe, *The Bloomsbury Review*

"Thomas Berry has demonstrated once again that he is one of the few great religious minds to be reckoned with."

Wes Jackson, president of the Land Institute

"Great Work indeed! Thomas Berry offers us the benefit of a lifetime of clear-headed, clear-hearted reflection. And by so doing he shows us where our task lies, shows us the particular test that we must face just as our ancestors faced their own great challenges. It's a work to stir the blood."

Bill McKibben, author of *The End of Nature*

"Thomas Berry is the bard of the new cosmology. He unerringly finds the mythic dimension and the moral significance behind the scientific facts."

Theodore Roszak, author of *The Voice of the Earth* and *Ecopsychology*

"How different American society might be if every high school student were exposed to the ideas contained in this book. One can only sigh in gratitude for this comprehensive and cautionary cultural history, and raise a cheer for those members of the human community already engaged in the truly Great Work."

Virginia Baron, *Parabola*

Also by Thomas Berry

The Historical Theory of Giambattista Vico

Buddhism

The Religions of India

The Dream of the Earth

Befriending the Earth (with Thomas Clarke)

The Universe Story (with Brian Swimme)

THOMAS BERRY

The GREAT WORK

Our Way into the Future

BELL TOWER NEW YORK

Some of the material in this book previously appeared in a different form.

Published by Bell Tower, New York, New York
Member of the Crown Publishing Group.
Originally published in hardcover by Bell Tower in 1999.

Random House, Inc. New York, Toronto, London, Sydney, Auckland
www.randomhouse.com

Bell Tower and colophon are registered trademarks of Random House, Inc.

Printed in the United States of America

Design by Barbara Sturman

Library of Congress Cataloging-in-Publication Data

Berry, Thomas Mary, 1914–
The great work : our way into the future / Thomas Berry. — 1st ed.
p. cm.
Includes bibliographical references.
1. Deep ecology—Philosophy. 2. Human ecology—Philosophy.
I. Title.
GE195.B47 1999
304.2—dc21 99-26350
CIP

ISBN 0–609–80499–5

10 9 8 7 6 5 4

To the children

To all the children

To the children who swim beneath

The waves of the sea, to those who live in

The soils of the Earth, to the children of the flowers

In the meadows and the trees in the forest, to

All those children who roam over the land

And the winged ones who fly with the winds,

To the human children too, that all the children

May go together into the future in the full

Diversity of their regional communities.

TO ALL THOSE WHO

HAVE ASSISTED ME THROUGH

THESE MANY YEARS,

I AM GRATEFUL.

Contents

Introduction

HUMAN PRESENCE ON THE PLANET EARTH IN THE OPENING years of the twenty-first century is the subject of this book. We need to understand where we are and how we got here. Once we are clear on these issues we can move forward with our historical destiny, to create a mutually enhancing mode of human dwelling on the planet Earth.

Just now we seem to be expecting some wonderworld to be attained through an ever-greater dedication to our sciences, technologies, and commercial projects. In the process, however, we are causing immense ruin in the world around us.

We might begin to think about our present life-situation by reflecting for a moment on the wonder of Earth, how it came to be the garden planet of the universe and what might be our human role in this context. To appreciate our immediate situation we might also develop a new intimacy with the North American continent. For we need the guidance and support of this continent as we find our way into the future.

The most basic and most disturbing commitment of the original European settlers was to conquer this continent and reduce it to human use. Because the exaltation of the human and the subjugation of the natural have been so excessive, we need to understand how the human community and the living forms of Earth might now become a life-giving presence to each other. We have already shaped the critical understanding and the appropriate technologies that can assist in establishing a beneficial human presence with the other components of this continent and also with the one great Earth community. We need only see that our human technologies are coherent with the ever-renewing technologies of the planet itself.

An indispensable resource in fulfillment of this task is the guidance of the indigenous peoples of this continent, for they have understood, better than we have understood, the integral relation of humans with this continent and with the entire natural world. In an earlier period we have been profoundly concerned with divine-human relations. In more recent centuries we have been concerned with interhuman relations. Our future destiny rests even more decisively on our capacity for intimacy in our human-Earth relations.

Of the institutions that should be guiding us into a viable future, the university has a special place because it teaches all those professions that control the human endeavor. In recent centuries the universities have supported an exploitation of the Earth by their teaching in the various professions, in the sciences, in engineering, law, education, and economics. Only in literature, poetry, music, art, and occasionally in religion and the biological sciences, has the natural world received the care that it deserves.

Our educational institutions need to see their purpose not as training personnel for exploiting the Earth but as guiding students toward an intimate relationship with the Earth. For it is the planet itself that brings us into being, sustains us in life, and delights us with its wonders. In this context we might consider the intellectual, political, and economic orientations that will enable us to fulfill the historical assignment before us—to establish a more viable way into the future.

As in creating some significant work the artist first experiences something akin to dream awareness that becomes clarified in the creative process itself, so we must first have a vision of the future sufficiently entrancing that it will sustain us in the transformation of the human project that is now in process. Such an entrancing vision we propose here as the Ecozoic Era, the period when humans would become a mutually beneficial presence on the Earth.

That future can exist only when we understand the universe as composed of subjects to be communed with, not as objects to be exploited. "Use" as our primary relationship with the planet must be abandoned. While there are critical issues in providing food, shelter, and livelihood to vast numbers of peoples, these issues themselves ultimately depend on our capacity to sustain the natural world so that the natural world can sustain us. All our sciences and technologies and all our social institutions become dysfunctional if the natural life systems cease to function.

Intimacy with the planet in its wonder and beauty and the full depth of its meaning is what enables an integral human relationship with the planet to function. It is the only possibility for humans to attain their true flourishing while honoring the other modes of earthly being. The fulfillment of the Earth community is to be caught up in the grandeur of existence itself and in admiration of those mysterious powers whence all this has emerged.

Nourishment of both the outer body and the inner spirit will be achieved in intimate association with each other or not at all. That we can now understand and work toward this fulfillment is the challenging future that opens up before us in these early years of the twenty-first century.

The
GREAT WORK

1

The GREAT WORK

HISTORY IS GOVERNED BY THOSE OVERARCHING MOVEMENTS that give shape and meaning to life by relating the human venture to the larger destinies of the universe. Creating such a movement might be called the Great Work of a people. There have been Great Works in the past: the Great Work of the classical Greek world with its understanding of the human mind and creation of the Western humanist tradition; the Great Work of Israel in articulating a new experience of the divine in human affairs; the Great Work of Rome in gathering the peoples of the Mediterranean world and of Western Europe into an ordered relation with one another. So too in the medieval period there was the task of giving a first shape to the Western world in its Christian form. The symbols of this Great Work were the medieval cathedrals rising so graciously into the heavens from the region of the old Frankish empire. There the divine and the human could be present to each other in some grand manner.

In India the Great Work was to lead human thought into spiritual experiences of time and eternity and their mutual presence to each other with a unique subtlety of expression. China created one of the most elegant and most human civilizations we have ever known as its Great Work. In America the Great Work of the First Peoples was to occupy this continent and establish an intimate rapport with the powers that brought this continent into existence in all its magnificence. They did this through their ceremonies such as the Great Thanksgiving ritual of the Iroquois, the sweat lodge and the vision quest of the Plains Indians, through the Chantways of the Navaho, and the Katsina rituals of the Hopi. Through these and a multitude of other aspects of the indigenous cultures of this continent, certain models were established of how humans become integral with the larger context of our existence here on the planet Earth.

While all of these efforts at fulfilling a Great Work have made significant contributions to the human venture, they were all limited in their fulfillment and bear the marks of their deeply human flaws and imperfections. Here in North America it is with a poignant feeling and foreboding concerning the future that we begin to realize that the European occupation of this continent, however admirable its intentions, has been flawed from the beginning in its assault on the indigenous peoples and its plundering of the land. Its most impressive achievements were establishing for the settlers a sense of personal rights, participatory governance, and religious freedom.

If there was also advancement of scientific insight and technological skills leading to relief from many of the ills and poverty of the European peoples, this advancement was accompanied by devastation of this continent in its natural florescence by the suppression of the way of life of its indigenous peoples and by communicating to them many previously unknown diseases, such as smallpox, tuberculosis, diphtheria, and measles. Although Europeans had developed a certain immunity to these diseases, they were consistently fatal to Indians, who had never known such diseases and had developed no immunities.

Meanwhile the incoming Europeans committed themselves to development of the new industrial age that was beginning to dominate human consciousness. New achievements in science, technology, industry, commerce, and finance had indeed brought the human community into a new age. Yet those who brought this new historical period into being saw only the bright side of these achievements. They had little comprehension of the devastation they were causing on this continent and throughout the planet, a devastation that finally led to an impasse in our relations with the natural world. Our commercial-industrial obsessions have disturbed the biosystems of this continent in a depth never known previously in the historical course of human affairs.

The Great Work now, as we move into a new millennium, is to carry out the transition from a period of human devastation of the Earth to a period when humans would be present to the planet in a mutually beneficial manner. This historical change is something more than the transition from the classical Roman period to the medieval period, or from the medieval period to modern times. Such a transition has no historical parallel since the geobiological transition that took place 67 million years ago when the period of the dinosaurs was terminated and a new biological age begun. So now we awaken to a period of extensive disarray in the biological structure and functioning of the planet.

Since we began to live in settled villages with agriculture and domestication of animals some ten thousand years ago, humans have put increased burdens upon the biosystems of the planet. These burdens were to some extent manageable because of the prodigality of nature, the limited number of humans, and their limited ability to disrupt the natural systems. In recent centuries, under the leadership of the Western world, largely with the resources, psychic energy, and inventiveness of the North American peoples, an industrial civilization has come into being with the power to plunder Earth in its deepest foundations, with awesome impact on its geological structure, its chemical constitution, and its living forms throughout the wide expanses of the land and the far reaches of the sea.

Some 25 billion tons of topsoil are now being lost each year with untold consequences to the food supply of future generations. Some of the most abundant species of marine life have become commercially extinct due to overexploitation by factory fishing vessels and the use of drift nets twenty to thirty miles long and twenty feet deep. If we consider the extinctions taking place in the rain forests of the southern regions of the planet with the other extinctions, we find that we are losing large numbers of species each year. Much more could be said concerning the impact of humans on the planet, the disturbance caused by the use of river systems for waste disposal, the pollution of the atmosphere by the burning of fossil fuels, and the radioactive waste consequent on our use of nuclear energy. All of this disturbance of the planet is leading to the terminal phase of the Cenozoic Era. Natural selection can no longer function as it has functioned in the past. Cultural selection is now a decisive force in determining the future of the biosystems of the Earth.

The deepest cause of the present devastation is found in a mode of consciousness that has established a radical discontinuity between the human and other modes of being and the bestowal of all rights on the humans. The other-than-human modes of being are seen as having no rights. They have reality and value only through their use by the human. In this context the other than human becomes totally vulnerable to exploitation by the human, an attitude that is shared by all four of the fundamental establishments that control the human realm: governments, corporations, universities, and religions—the political, economic, intellectual, and religious establishments. All four are committed consciously or unconsciously to a radical discontinuity between the human and the nonhuman.

In reality there is a single integral community of the Earth that includes all its component members whether human or other than human. In this community every being has its own role to fulfill, its own dignity, its inner spontaneity. Every being has its own voice. Every being declares itself to the entire universe. Every being enters into communion with other beings. This capacity for relatedness, for presence to other beings, for spontaneity in action, is a capacity possessed by every mode of being throughout the entire universe.

So too every being has rights to be recognized and revered. Trees have tree rights, insects have insect rights, rivers have river rights, mountains have mountain rights. So too with the entire range of beings throughout the universe. All rights are limited and relative. So too with humans. We have human rights. We have rights to the nourishment and shelter we need. We have rights to habitat. But we have no rights to deprive other species of their proper habitat. We have no rights to interfere with their migration routes. We have no rights to disturb the basic functioning of the biosystems of the planet. We cannot own the Earth or any part of the Earth in any absolute manner. We own property in accord with the well-being of the property and for the benefit of the larger community as well as ourselves.

A sense of the continent being here primarily for our use has been developing throughout the past few centuries. We proceeded with our destruction of the forests until the terminal phase of the twentieth century, when we found that we had cut down over 95 percent of the primordial forests of this continent. With the new technologies that emerged in the last half of the nineteenth century and the automobile industry that developed in the early twentieth century, industrialization achieved a new virulence. Roadways, superhighways, parking lots, shopping centers, malls, and housing developments took over. Suburban living became normative for the good life. This was also the time when the number of free-flowing rivers began to decline. The great dams were built on the Colorado, the Snake, and especially the Columbia rivers.

Yet this was also the time when resistance began. The increasing threat to the natural life-systems of the continent awakened the sense of need for grandeur in the natural world if any truly human development was to continue in our cultural traditions. This new awareness began in the nineteenth century with such persons as Henry David Thoreau, John Muir, John Burroughs, and George Perkin Marsh; with John Wesley Powell and Frederick Law Olmstead; also with artists, especially Thomas Cole, Frederick Edwin Church, and Albert Bierstadt of the Hudson River School.

To the work of naturalists and artists was added the work of the conservationists in the political realm. These leaders brought about the preservation of Yellowstone National Park in 1872, the first wilderness area anywhere on Earth to be officially set aside for preservation in perpetuity. Later, in 1885, New York State established the Adirondack Forest Preserve, a region to be kept forever in its wilderness condition. In 1890 Yosemite National Park was established in California. In that same period the first voluntary associations were formed to foster a deepened appreciation of the natural world. The Audubon Society, founded in 1886, was concerned primarily with appreciation of the various bird species. The Sierra Club was founded in 1892 and the Wilderness Society in 1924. Both sought to create a more intimate relationship between the human community and the wild world about us.

These various groups were the beginning. The larger dimensions of what was happening could not have been known to those living in the nineteenth century. They could not have foreseen the petroleum industry, the automobile age, the damming of the rivers, the emptying of the marine life of the oceans, the radioactive waste. Yet they knew that something was wrong at a profound level. Some, such as John Muir, were deeply disturbed. When the decision was made to build a dam to enclose Hetch-Hetchy Valley as a reservoir for the city of San Francisco, he considered it the unnecessary destruction of one of the most sacred shrines in the natural world, a shrine that fulfilled some of the deepest emotional, imaginative, and intellectual needs of the human soul. "Dam Hetch-Hetchy! As well dam for water-tanks the people's cathedrals and churches, for no holier temple has ever been consecrated by the heart of man" (Teale, p. 320).

Throughout the twentieth century the situation has worsened decade by decade with relentless commitment to making profit by ruining the planet for the uncertain benefit of the human. The great corporations have joined together so that a few establishments now control vast regions of the Earth. Assets of a few transnational corporations begin to rise toward the trillion-dollar range. Now, in these closing years of the twentieth century, we find a growing concern for our responsibility to the generations who will live in the twenty-first century.

Perhaps the most valuable heritage we can provide for future generations is some sense of the Great Work that is before them of moving the human project from its devastating exploitation to a benign presence. We need to give them some indication of how the next generation can fulfill this work in an effective manner. For the success or failure of any historical age is the extent to which those living at that time have fulfilled the special role that history has imposed upon them. No age lives completely unto itself. Each age has only what it receives from the prior generation. Just now we have abundant evidence that the various species of life, the mountains and rivers, and even the vast ocean itself, which once we thought beyond serious impact from humans, will survive only in their damaged integrity.

The Great Work before us, the task of moving modern industrial civilization from its present devastating influence on the Earth to a more benign mode of presence, is not a role that we have chosen. It is a role given to us, beyond any consultation with ourselves. We did not choose. We were chosen by some power beyond ourselves for this historical task. We do not choose the moment of our birth, who our parents will be, our particular culture or the historical moment when we will be born. We do not choose the status of spiritual insight or political or economic conditions that will be the context of our lives. We are, as it were, thrown into existence with a challenge and a role that is beyond any personal choice. The nobility of our lives, however, depends upon the manner in which we come to understand and fulfill our assigned role.

Yet we must believe that those powers that assign our role must in that same act bestow upon us the ability to fulfill this role. We must believe that we are cared for and guided by these same powers that bring us into being.

Our own special role, which we will hand on to our children, is that of managing the arduous transition from the terminal Cenozoic to the emerging Ecozoic Era, the period when humans will be present to the planet as participating members of the comprehensive Earth community. This is our Great Work and the work of our children, just as Europeans in the twelfth and thirteenth centuries were given the role of bringing a new cultural age out of the difficulties and strife of that long period from the sixth through the eleventh centuries. At this time, the grandeur of the classical period had dissolved, the cities of Europe had declined, and life in all its physical and cultural aspects was carried on in the great castles and monasteries to constitute what came to be known as the manorial period in European history.

In the ninth and tenth centuries the Normans were invading the nascent culture of Europe from the north, the Magyars were moving in from the east, and the Muslims were advancing in Spain. Western civilization was situated in a very limited region under siege. In response to this threatening situation, medieval Europe toward the end of the eleventh century began the crusading wars that united the nations of Europe and for two centuries engaged them in an eastward drive toward Jerusalem and the conquest of the Holy Land.

This period might be considered the beginning of the historical drive that has led European peoples in their quest for religious, cultural, political, and economic conquest of the world. This movement was continued through the period of discovery and control over the planet into our own times when the Western presence culminates politically in the United Nations and economically in such establishments as the World Bank, the International Monetary Fund, the World Trade Organization, and the World Business Council for Sustainable Development. We might even interpret this Western drive toward limitless dominion in all its forms as leading eventually to the drive toward human dominion over the natural world.

The immediate achievement, however, of the thirteenth century was the creation of the first integration of what became Western civilization. In this century new and dazzling achievements took place in the arts, in architecture, in speculative thinking, in literature. By raising up the medieval cathedrals a new and original architecture was created. In these soaring structures an artistic daring and refinement was manifest that has been equaled only in rare moments in the larger history of civilizations. This was also the period of Francis, the poor man of Assisi who established in Western civilization both the spiritual ideal of detachment from Earthly possessions and an intimacy with the natural world. It was also the period of Thomas Aquinas, who originated Aristotelian studies, especially in the field of cosmology, in medieval Christian civilization. Within this context Thomas reinterpreted the entire range of Western theological thought. As the philosopher Alfred North Whitehead has noted, this was the time when the Western mind took on that critical keenness and reasoning process that made our modern scientific thought processes possible. In literature the incomparable Dante Alighieri produced his *Commedia* in the early fourteenth century, a time when Giotto was already beginning, with Cimabue, the great period of Italian painting.

The importance of recalling these shaping forces in the narrative account of Western civilization is that they arose as a response to the Dark Ages from the sixth- through eleventh-centuries in Europe. We need to recall that in these and in so many other instances the dark periods of history are the creative periods; for these are the times when new ideas, arts, and institutions can be brought into being at the most basic level. Just as the brilliant period of medieval civilization arose out of these earlier conditions, so we might recall the period in China when, in the third century, the tribal invasions from the northwest had broken down the rule of the Han dynasty and for several centuries brought about a disunity throughout the empire. Yet this period of dissolution was also the period of Buddhist monks and Confucian scholars and artists who gave expression to new visions and new thoughts at the deepest levels of human consciousness. Scholars carrying the Taoist and Confucian traditions would later inspire literary figures such as Li Po, Tu Fu, and Po Chü-i in the T'ang period of the eighth century. Following the T'ang period the Sung period of the tenth through fourteenth centuries would bring forth such masterful interpretations of traditional Chinese thought as those presented by Chou

Tun-i and Chu Hsi. Artists such as Ma Yüan and Hsia Kuei of the twelfth century and poets such as Su Tung-p'o would complete this creative period in the cultural history of China. These are some of the persons who enabled the Chinese to survive as a people and as a culture and to discover new expressions of themselves after this long period of threats to their survival.

We must consider ourselves in these early years of the twenty-first century as also experiencing a threatening historical situation, although our situation is ultimately beyond comparison with any former period in Europe or in Asia. For those peoples were dealing with human adjustment to disturbances of human life patterns. They were not dealing with the disruption and even the termination of a geobiological period that had governed the functioning of the planet for some 67 million years. They were not dealing with anything comparable to the toxics in the air, the water, and the soil, or with the immense volume of chemicals dispersed throughout the planet. Nor were they dealing with the extinction of species or the altering of the climate on the scale of our present concern.

Yet we can be inspired by their example, their courage, and even by their teachings. For we are heirs to an immense intellectual heritage, to the wisdom traditions whereby they were able to fulfill the Great Work of their times. These traditions are not the transient thoughts or immediate insights of journalists concerned with the daily course of human affairs; these are expressions in human form of the principles guiding human life within the very structure and functioning of the universe itself.

We might observe here that the Great Work of a people is the work of all the people. No one is exempt. Each of us has our individual life pattern and responsibilities. Yet beyond these concerns each person in and through their personal work assists in the Great Work. Personal work needs to be aligned with the Great Work. This can be seen in the medieval period as the basic patterns of personal life and craft skills were aligned within the larger work of the civilizational effort. While this alignment is more difficult in these times it must remain an ideal to be sought.

We cannot doubt that we too have been given the intellectual vision, the spiritual insight, and even the physical resources we need for carrying out the transition that is demanded of these times, transition from the period when humans were a disruptive force on the planet Earth to the period when humans become present to the planet in a manner that is mutually enhancing.

2

The MEADOW *across the* CREEK

MY OWN UNDERSTANDING OF THE GREAT WORK BEGAN WHEN I was quite young. At the time I was some eleven years old. My family was moving from a more settled part of a small southern town out to the edge of town where the new house was being built. The house, not yet finished, was situated on a slight incline. Down below was a small creek and there across the creek was a meadow. It was an early afternoon in late May when I first wandered down the incline, crossed the creek, and looked out over the scene.

The field was covered with white lilies rising above the thick grass. A magic moment, this experience gave to my life something that seems to explain my thinking at a more profound level than almost any other experience I can remember. It was not only the lilies. It was the singing of the crickets and the woodlands in the distance and the clouds in a clear sky. It was not something conscious that happened just then. I went on about my life as any young person might do.

Perhaps it was not simply this moment that made such a deep impression upon me. Perhaps it was a sensitivity that was developed throughout my childhood. Yet as the years pass this moment returns to me, and whenever I think about my basic life attitude and the whole trend of my mind and the causes to which I have given my efforts, I seem to come back to this moment and the impact it has had on my feeling for what is real and worthwhile in life.

This early experience, it seems, has become normative for me throughout the entire range of my thinking. Whatever preserves and enhances this meadow in the natural cycles of its transformation is good; whatever opposes this meadow or negates it is not good. My life orientation is that simple. It is also that pervasive. It applies in economics and political orientation as well as in education and religion.

That is good in economics which fosters the natural processes of this meadow. That is not good in economics which diminishes the capacity of this meadow to renew itself each spring and to provide a setting in which crickets can sing and birds can feed. Such meadows, I later learned, are themselves in a continuing process of transformation. Yet these evolving biosystems deserve the opportunity to be themselves and to express their own inner qualities. As in economics, so in jurisprudence and law and political affairs—what is good recognizes the rights of this meadow and the creek and the woodlands beyond to exist and flourish in their ever-renewing seasonal expression even while larger processes shape the bioregion in its sequence of transformations.

Religion too, it seems to me, takes its origin here in the deep mystery of this setting. The more a person thinks of the infinite number of interrelated activities that take place here, the more mysterious it all becomes. The more meaning a person finds in the Maytime blooming of the lilies, the more awestruck a person might be in simply looking out over this little patch of meadowland. It has none of the majesty of the Appalachian or the western mountains, none of the immensity or the power of the oceans, nor even the harsh magnificence of desert country. Yet in this little meadow the magnificence of life as celebration is manifested in a manner as profound and as impressive as any other place I have known in these past many years.

It seems to me many people had such experiences before we entered into an industrial way of life. The universe, as manifestation of some primordial grandeur, was recognized as the ultimate referent in any human understanding of the wonderful yet fearsome world about us. Every being achieved its full identity by its alignment with the universe itself. With indigenous peoples of the North American continent every formal activity was first situated in relation to the six directions of the universe, the four cardinal directions combined with the heavens above and the Earth below. Only thus could any human activity be fully validated.

The universe was the world of meaning in these earlier times, the basic referent in social order, in economic survival, in the healing of illness. In that wide ambience the Muses dwelled, whence came the inspiration of poetry and art and music. The drum, heartbeat of the universe itself, established the rhythm of dance, whereby humans entered into the entrancing movement of the natural world. The numinous dimension of the universe impressed itself upon the mind through the vastness of the heavens and the power revealed in the thunder and lightning, as well as through the springtime renewal of life after the desolation of winter. Then too the general helplessness of the human before all the threats to survival revealed the intimate dependence of the human on the integral functioning of things. That the human had such intimate rapport with the surrounding universe was possible only because the universe itself had a prior intimate rapport with the human as the maternal source from whence humans come into being and are sustained in existence.

This experience we observe even now in the indigenous peoples of the world. They live in a universe, in a cosmological order, whereas we, the peoples of the industrial world, no longer live in a universe. We in North America live in a political world, a nation, a business world, an economic order, a cultural tradition, a Disney dreamland. We live in cities, in a world of concrete and steel, of wheels and wires, a world of unending work. We seldom see the stars at night or the planets or the moon. Even in the day we do not experience the sun in any immediate or meaningful manner. Summer and winter are the same inside the mall. Ours is a world of highways, parking lots, shopping centers. We read books written with a strangely contrived human alphabet. We no longer read the Book of Nature.

While we have more scientific knowledge of the universe than any people ever had, it is not the type of knowledge that leads to an intimate presence within a meaningful universe. The various phenomena of nature are not spirit presences. We no longer read the book of the universe. We have extensive contact with the natural world through photographs and television presentations. But as Saint Augustine remarked long ago, a picture of food does not nourish us. Our world of human meaning is no longer coordinated with the meaning of our surroundings. We have disengaged from that profound interaction with our environment that is inherent in our nature. Our children no longer learn how to read the great Book of Nature from their own direct experience or how to interact creatively with the seasonal transformations of the planet. They seldom learn where their water comes from or where it goes. We no longer coordinate our human celebrations with the great liturgy of the heavens.

So completely are we at odds with the planet that brought us into being that we have become strange beings indeed. We dedicate enormous talent and knowledge and research in developing a human order disengaged from and even predatory on the very sources from whence we came and upon which we depend every moment of our existence. We initiate our children into an economic order based on exploitation of the natural life systems of the planet. To achieve this attitude we must first make our children unfeeling in their relation with the natural world. This occurs quite simply since we ourselves have become insensitive toward the natural world and do not realize just what we are doing. Yet if we observe our children closely in their early years we see how they are instinctively attracted to profound experiences of the natural world. We also see additional stresses, emotional disruptions, and learning disabilities that seem to originate in the toxic environment and processed food that we provide for them.

A primary concern for the peoples of this continent must be to recover an integral relation with the universe, the planet Earth, and the North American continent. While a new alignment of our government, our institutions, and our professions with the continent in its deep structure and functioning cannot be achieved immediately, a beginning can be made through our educational programs. Especially in the early grades of elementary school, new developments are possible. Such was the insight of the educator Maria Montessori in the earlier part of the twentieth century.

In speaking about the education of the six-year-old child, she notes in her book, *To Educate the Human Potential*, that only when the child is able to identify its own center with the center of the universe does education really begin. For the universe, she says, "is an imposing reality." It is "an answer to all questions." "We shall walk together on this path of life, for all things are part of the universe, and are connected with each other to form one whole unity." This comprehensive context enables "the mind of the child to become centered, to stop wandering in an aimless quest for knowledge." She observes how this experience of the universe creates in children admiration and wonder, how this enables children to unify their thinking. In this manner children learn how all things are related and how the relationship of things to one another is so close that "no matter what we touch, an atom, or a cell, we cannot explain it without knowledge of the wide universe" (Montessori, p. 6).

The difficulty is that with the rise of the modern sciences we began to think of the universe as a collection of objects rather than as a communion of subjects. We frequently discuss the loss of the interior spirit world of the human mind with the rise of the modern mechanistic sciences. The more significant realization, however, is that we have lost the universe itself. We have achieved extensive control over the mechanistic and even the biological functioning of the natural world, but this control has not always had beneficial consequences. We have not only controlled the planet in much of its basic functioning, we have, to an extensive degree, extinguished the life systems themselves. We have silenced too many of those wonderful voices of the universe that spoke to us of the grand mysteries of existence.

We no longer hear the voice of the rivers, the mountains, or the sea. The trees and meadows are no longer intimate modes of spirit presence. The world about us has become an "it" rather than a "thou," as was noted by the distinguished archaeologist Henri Frankfort in *Before Philosophy* (p. 26). We continue to make music, write poetry, and do our painting and sculpture and architecture, but these activities easily become aesthetic expressions simply of the human. They lose the intimacy and radiance and awesome qualities of the universe. We have, in the accepted universe of these times, little capacity for participating in the mysteries that were celebrated in the earlier literary and artistic and religious modes of expression. For we cannot live in the universe in which these celebrations took place. We can only look on, as it were, as at something unreal.

Yet the universe is so bound into the aesthetic experience, into poetry, music, art, and dance, that we cannot entirely avoid the implicit dimensions of the natural world. This is true even when we think of art as "representational" or "impressionist" or "expressionist" or as "personal statement." However we think of our art or literature, its power is there in the wonder communicated most directly by the meadow or the mountains, by the sea or by the stars in the night.

Of special significance is our capacity for celebration, which inevitably brings us into the rituals that provide the coordination of human affairs with the great liturgy of the universe. Our national holidays, political events, and heroic human deeds are all quite worthy of celebration, but ultimately, unless they are associated with some more comprehensive level of meaning, they tend toward the affected, the emotional, and the ephemeral. Here we might note, however, that in the political and legal orders we have never been able to give up our invocation of the more sublime dimensions of the universe to witness the truth of what we say. This we observe in the oaths that we swear in inaugural ceremonies, in official documents, and in court trials. We still have an instinctive awe and reverence and even a certain fear of the larger world that always lies outside the range of our human controls.

Even when we recognize the spirit world beyond the human we make everything referent to the human as the ultimate source of meaning and of value, although this way of thinking has led to catastrophe for ourselves as well as a multitude of other beings. Yet in recent times we begin to recognize that the universe, in the phenomenal order, is the only self-referent mode of being. All other modes of being, including the human, in their existence and in their functioning, are universe-referent. This relation with the universe has been recognized through the centuries in the rituals of the various traditions.

From Paleolithic times humans have coordinated their ritual celebrations with the transformation moments of the natural world. Ultimately the universe, throughout its vast extent in space and its sequence of transformations in time, was seen as a single, multiform celebratory expression. No other explanation seems possible for the world that we see around us. The birds fly and sing; they build their nests and raise their young. The flowers blossom. The rains nourish every living being. The tides flow back and forth. The seasons succeed each other in an entrancing sequence. Each of the events in the natural world is a poem, a painting, a drama, a celebration.

Dawn and sunset are the mystical moments of the diurnal cycle, the moments when the numinous dimension of the universe reveals itself with special intimacy. Individually and in their relations with each other these are moments when the high meaning of existence is experienced. Whether in the gatherings of indigenous peoples in their tribal setting or in the more elaborate temples and cathedrals and spiritual centers throughout the Earth these moments are celebrated with special observances. So too in the yearly cycle spring is celebrated as the time for renewal of the human in a proper alignment with the universal order of things.

The proposal has been made that no effective restoration of a viable mode of human presence on the planet will take place until such intimate human rapport with the Earth community and the entire functioning of the universe is reestablished on an extensive scale. Until this is done the alienation of the human will continue despite the heroic efforts being made toward a more benign mode of human activity in relation to the Earth. The present is not a time for desperation but for hopeful activity. This we discover in the firm reassertion of traditional thought and rituals that we can observe with the indigenous peoples of this continent. This we find in the teachings of Black Elk and in the resurgence of the Sundance ritual with the Crow Indians. In the writings of Scott Momaday, the inspiration of Lame Deer, the guidance of Oren Lyons, the poetry of Joy Harjo, the essays of Linda Hogan, and the insight of Vine Deloria, we find a renewal of indigenous thought and a critical response to the traditional religious and scientific modes of Western thought. In each of these we find an intimate presence of the human venture with the great cosmic liturgy of the natural world.

In accord with indigenous modes of thinking throughout the world we might give a certain emphasis to the need to understand the universe primarily as celebration. While the universe celebrates itself in every mode of being, the human might be identified as that being in whom the universe celebrates itself and its numinous origins in a special mode of conscious self-awareness. Spontaneous forms of community ritual, such as the Council of All Beings inaugurated by John Seed, have already been developed. So too the ritual programs of Joanna Macy, the solstice celebrations of Paul Winter, and the seasonal festivals of Genesis Farm all give promise of a future with the understanding, the power, the aesthetic grandeur, and the emotional fulfillment needed to heal the damage that has already been wrought on the planet and to shape for Earth a viable future, a future with the entrancing qualities needed to endure the difficulties to be encountered and to evoke the creativity needed.

Here I would suggest that the work before us is the task not simply of ourselves but of the entire planet and all its component members. While the damage that has been done is immediately the work of humans, the healing cannot be the work simply of humans any more than the illness of one organ of the body can be healed through the efforts of that one organ. Every member of the body must bring about the healing. So now the entire universe is involved in the healing of the damaged Earth in the light and warmth of the sun.

As Earth is, in a sense, a magic planet in the exquisite presence of its diverse members to one another, so this movement into the future must in some manner be brought about in ways that are ineffable to the human mind. We might think of a viable future for the planet less as the result of some scientific insight or as dependent on some socioeconomic arrangement than as participation in a symphony or as renewed presence to some numinous presence manifested in the wonderworld about us. This was perhaps something I vaguely experienced in that first view of the lilies blooming in the meadow across the creek.

3

The EARTH STORY

OUR GREAT WORK, OUR HISTORICAL ROLE IN ITS DEEPER significance, has to do with a new understanding of the planet Earth: this radiant blue-white planet hanging in the sky, twirling upon its axis in the light of the sun each day, swinging in its solar orbit each year. Seven continents rise out of the great world ocean. The polar regions appear as vast stretches of snow. The Sierra Nevada along the western edge of the Americas, the Alps in Europe, the Himalayas in Nepal, the T'ien Shan in China, Kilimanjaro in Africa—these give to the continents a foreboding majesty. The rivers flow down from the mountains across the continents into the sea. Rain forests girdle the planet in its equatorial regions. Such vistas create an overwhelming impression whether we look down from the heavens or across the landscape and up at the sky with its sun and clouds in the day and its moon and stars in the night.

We seldom think about the Earth itself in its distinctive aspects; we are enclosed so intimately within its fields and woodlands or lost amid the commercial frenzy of our cities. We do speak about nature, the world, creation, the environment, the universe, even when the planet Earth is foremost in our thoughts. Yet only in recent times have we experienced the Earth in its full spherical contours.

The more we learn about the Earth the more clearly we see it as a privileged planet, a creation and the homeland of a multitude of living beings. Recently we have come to know the Earth within the context of a more comprehensive knowledge of the universe itself. Through our observational sciences we begin to understand just how the Earth was born out of the larger processes of the universe, how life appeared on the land, and more recently how we ourselves emerged into being. But if we have such scientific knowledge, we are often lacking in any deep feeling for the mystique of the Earth or any depth of understanding. We think of the Earth more as the background for economic purposes or as the object of scientific research rather than as a world of wonder, magnificence, and mystery for the unending delight of the human mind and imagination.

Earlier, Earth was a more intimate reality than it is at present. Animals and humans were relatives. This relationship found visible expression in the totemic carvings discovered throughout the world. The powers of the universe were grandfathers and grandmothers. A pervasive rapport with the spirit powers of the natural world was developed. Ritual enabled humans to enter into the grand liturgy of the universe itself. Seasonal renewal ceremonies brought humans into the rhythms of the solar and lunar cycle. Architectural structures were set on coordinates identified with the position of the heavenly bodies, something we seldom think about anymore.

This was a period of wonder and creativity that was to shape the human project until our times. Without confusing this earlier historical period with a mythical paradise, there did exist at moments a remarkable continuity throughout the various realms of existence. Human activities were integral with the larger community and its functioning. Every being possessed its own life principle, its own mode of self-expression, its own voice. Humans, animals, and plants and all natural phenomena were integral components of the larger

Earth community. As Henri Frankfort mentions in *Beyond Philosophy,* "Natural phenomena were regularly conceived in terms of human experience and human experience was conceived in terms of cosmic events" (Frankfort, p. 12).

The human and the cosmic responded to each other most extensively in the Chinese world. Human activities throughout the year were coordinated with the cycle of the seasons. So we find in the *Li Chi,* the Book of Ritual, prescriptions indicating that the robes of the emperor, the palace rooms in which he lived, the music and the ceremonies were all carefully coordinated with the seasons. If springtime music was played in the autumn then the cosmic order was considered to be disrupted (quoted in Legge, p. 291).

As indicated by Wang Yang-ming in his *Questions on the Great Learning,* the supreme achievement of the human personality in this context was to experience one's self as "one body" with "heaven and Earth and the myriad things" (quoted in deBary and Bloom, pp. 845–46). In the vast creative processes of the universe the human was "a third along with heaven and Earth" as a primordial force shaping the entire order of things.

Continuity of the human with the natural world in a single sacred community can be appreciated in the experience of Black Elk, a Lakota Sioux Indian, as narrated in his life story, *Black Elk Speaks.* When he was nine years old he experienced an elaborate vision culminating in a vast cosmic dance evoked by the song of the black stallion seen in the heavens (Neihardt, chap. 2). A sense of the sacred dimension of the universe is evident here, a type of awareness of the natural world that seems to be less available from our modern Western religions.

Earlier such intimacy did exist in the mystical tradition of persons such as Hildegard of Bingen, Richard of St. Victor, Meister Eckhart, and John of the Cross. From an early period Christians adopted a liturgy that carefully observed the correspondences of human praise with the numinous moments of the dawn and sunset and with the transitions of the various seasons of the year. This liturgy was carried out most faithfully in the Benedictine and Cistercian monasteries of Europe up through the medieval period. The social order was itself governed by this basic rhythm of life. Holy days were the original holidays, as the words themselves indicate.

We lose our intimacy with the natural world once we take on a secular life attitude. The natural world becomes vulnerable to assault by humans. Although the Christian world did have this commitment to the cosmological order, it had an even deeper commitment to the historical order. This historical orientation is ultimately what made our Western world so powerful in its political and economic dominance throughout the world and yet so vulnerable to the loss of its intimate concern for the natural world. It also led to the concept of the natural world as being there primarily for "use" rather than as manifestation of some numinous presence.

Now, after these centuries of experiencing the planet as being a collection of objects for scientific analysis and commercial use, we must ask: where can we find the resources for a reevaluation of our activities? How can we obtain the psychic energies needed to disengage from our plundering industrial economy? We might begin with our basic sense of reality as this exists at present. Our sense of reality cannot be simply the mythic worlds of the past, nor can it be limited to traditions that exist in a spatial mode of consciousness. Whatever be the case with other societies and other times we function through our observational sciences, in the context of a developmental universe that has, within the phenomenal world, its own self-organizing powers.

For our sense of reality three commitments are basic: to observational science, to a developmental universe, to an inner self-organizing capacity. We cannot do without our earlier experiences of the numinous presence manifested in the great Cosmic Liturgy. We cannot do without our humanistic traditions, our art and poetry and literature. But these traditions cannot themselves, simply with their own powers, do what needs to be done. These earlier experiences and accomplishments were dealing with other issues, providing guidance for different worlds than the world of the early twenty-first century. To meet the current environmental challenge they too need to be transformed within the context of an emergent universe.

Nor can the former accusations against the materialism of our scientific endeavor be accepted. Our observational sciences presently have moved beyond the mechanistic understanding of a so-called objective world as it was known in the past few centuries of Newtonian physics. We know now that there is a subjectivity in all our knowledge, that we ourselves, precisely as intelligent beings, activate one of the deepest dimensions of the universe. If formerly we knew by downward reduction processes that considered the particle as the reality and the wholes as derivative, we now recognize that it is even more important that we integrate upward, because we cannot know particles and their power until we see the wholes that they bring into being. If we know the elements simply in their isolated individual reality we have only minimal knowledge of what they really are. To understand atoms we must see these elements in their central role in molecules, megamolecules, in cellular life, organic life, even in intellectual perception, since atomic structures in a transformed context live and function in the wide display of all the gorgeous plants and animals of the Earth as well as in the most profound intellectual, emotional, and spiritual experiences of the human.

There is a spiritual capacity in carbon as there is a carbon component functioning in our highest spiritual experience. If some scientists consider that all this is merely a material process, then what they call matter, I call mind, soul, spirit, or consciousness. Possibly it is a question of terminology, since scientists too on occasion use terms that express awe and mystery. Most often, perhaps, they use the expression that some of the natural forms they encounter seem to be "telling them something."

It seems best to consider that mind and matter are two dimensions of the single reality that comes into being in an immense diversity of expression throughout the universe by some self-organizing process. We begin to appreciate this reality in the wonders of its achievements, although it will remain forever beyond expression in any scientific formulation or humanly constructed equation.

The second aspect of our present knowledge is that the universe is revealed to us as irreversible emergent process. We no longer live simply in a spatial mode of consciousness where time is experienced as a seasonal renewing sequence of realities that keep their basic identity in accord with the Platonic archetypal world. We now live not so much in a *cosmos* as in a *cosmogenesis*; that is, a universe ever coming into being through an irreversible sequence of transformations moving, in the larger arc of its development, from a lesser to a great order of complexity and from a lesser to great consciousness.

The third foundation for appreciating our own times is to recognize that there exists at every level a basic tendency toward self-organization. This we find at the physical level, at the biological level, and at the level of reflexive-consciousness. While the ancients had much more highly developed sensitivities regarding the natural world in its numinous aspects and in its inner spontaneities, we are not without our own resources that, properly appreciated, can lead to our own mode of intimacy with the natural world, and even to a renewal of the Earth in the new ecological community. If for a while we lost the poetry of the universe, this loss was significantly changed when the astronauts came home stunned with the immensity and beauty of what they had experienced. Especially overwhelming was their view of the planet Earth from the regions of the moon, almost 200,000 miles distant. A new poetic splendor suddenly appeared in their writings.

The astronaut Edgar Mitchell tells us that he had an amazing experience when he looked out at Earth from outer space and saw "this blue-and-white planet floating there," then saw the sun set "in the background of the very deep black and velvety cosmos." He was overcome with immersion in an awareness that there was "a purposefulness of flow, of energy, of time, of space in the cosmos" beyond any previous experience that he ever had (Kelley, p. 138).

This sensitive experience of the universe and of the Earth leads us further back to appreciation of the ten billion years required for the universe to bring the Earth into existence and another 4.6 billion years for the Earth to shape itself in such splendor. For our present Earth is not the Earth as it always was and always will be. It is the Earth at a highly developed phase in its continuing emergence. We need to see the Earth in its sequence of transformations as so many movements in a musical composition. The sequence of events that emerge in time needs to be understood simultaneously, as in music: the earlier notes are gone when the later notes are played, but the musical phrase, indeed the entire symphony, needs to be heard simultaneously. We do not fully understand the opening notes until the later notes are heard. Each new theme alters the meaning of the earlier themes and the entire composition. The opening theme resonates throughout all the later parts of the piece.

So too the origin moment of the universe presents us with an amazing process that we begin to appreciate as a mystery unfolding through the ages. The flaring forth of the primordial energy carried within itself all that would ever happen in the long series of transformations that would bring the universe into its present mode of being. The origin moment of the universe was the implicate form of the present as the present is the explicate form of the origin moment. The primordial emergence was the beginning of the Earth story, as well as the beginning of the personal story of each of us, since the story of the universe is the story of each individual being in the universe. Indeed the reality inherent in the original flaring forth could not be known until the shaping forces held in this process had brought forth the galaxies, the Earth, the multitude of living species, and the reflection of the universe on itself in human intelligence.

After the origin moment a sequence of other transformation moments took place, the shaping of the first generations of stars within their various galaxies, then the collapse of one of these stars into a vast dispersion of fragments throughout the realms of space. The energy of this supernova moment brought into being the entire array of elements. These elements in turn made possible the future developments on the planet Earth, for indeed the appearance of life needed the broad spectrum of elements for its full expression.

Our own solar system with its nine planets became possible at this moment. A gravitational center of attraction gathered the scattered fragments of the former star into this new star, our sun, and surrounding this star with its ninefold arrangement of planets. Within this context, the planet Earth began its distinctive self-expression, a groping toward its unknowable future, yet carrying within itself a tendency toward an ever greater differentiation, a deepening spontaneity, an ever more intimate self-bonding of its component parts. Such a wonderment comes over us as we reflect on Earth finding its proper distance from the sun so that it would be neither too hot nor too cold, shaping its radius so that it would be neither too large (and thus make Earth more gaseous like Jupiter) nor too small (and thus make Earth more rock-bound like Mars). Then the moon must be situated with such precision, that it would neither be so close that the tides would overwhelm the continents, nor so distant that the seas would be stagnant.

The radioactive elements within Earth provided the heat for the volcanic explosions leading to the atmosphere and the seas and raising the continents above the waters. Profound mysteries were taking place all this while, but nothing so mysterious as setting into place the conditions required for the emergence of life and the human form of consciousness. The 3.4 billion-year story of life is so integral with the story of Earth in its basic structure that we cannot properly think of the Earth as first taking shape in its full physical form and then life emerging somehow within this context.

Earth as we know it came into being through its four great components: land, water, air, and life, all interacting in the light and energy of the sun. Although there was a sequence in the formation of the land sphere, the atmosphere, the water sphere, and the life sphere, these have so interacted with one another in the shaping of the Earth that we must somehow think of these as all present to one another and interacting from the beginning.

Although there was a primordial atmosphere and sea and land, these were so transformed by life development that we might think of the Earth as primarily a life process. We do need to tell in sequence the story of the physical shaping of the Earth and the primary form of the atmosphere and the seas with their chemical components and then consider the procaryotic cell and invention of photosynthesis, the eucaryotic cell, and respiration, then the more elaborate expression of all these modes of Earth development. We must constantly realize that each stage of development was the consequence of a single process at work, a process that came to a new phase of its development in the human mode of consciousness. This unity of the universe was more easily appreciated in classical times when Plato in his *Timaeus* proposes the idea of a world-soul that gives a living unity to the entire universe. This idea of a world-soul, an *anima mundi*, continued in the European world until the seventeenth century with the Cambridge Platonists: Henry More, Richard Cumberland, and Ralph Cudworth.

Of more immediate significance to ourselves in this telling the story of the Earth is the sequence of life developments that has emerged in these past 600 million years, the time generally presented in terms of the Paleozoic (from 570 million years before present to 240 myp), the Mesozoic (240 myp until 65 myp) and the Cenozoic (65 myp). While it would be useful to discuss the earlier biological eras, it is the Cenozoic that is of most interest to us. This is the era when our world took shape. While many of the distinctive life-forms of the Cenozoic were already present in the earlier Mesozoic Era, they attained their full development in the Cenozoic. This is the era when the flowers came forth in all their gorgeous colors and fantastic shapes. It is the period of the great deciduous trees in the temperate zones and of the tropical rain forests in the equatorial region. The Cenozoic is the special time of the birds in all the variety of their forms and colors and songs and mating rituals. Above all it is the era of the mammals. The varied multitude of living species, possibly twenty million, came into their greatest splendor in this era. We will never know these species fully since many have come and gone in the natural process of evolutionary change. Now we ourselves are extinguishing species in a volume and with a rapidity far beyond any former natural processes of extinction since the beginning of the Cenozoic Era.

The late Cenozoic was a wildly creative period of inspired fantasy and extravagant play. It was a supremely lyrical moment when humans emerged on the scene, quietly, somewhere on the edge of the savanna in northeast Africa. From here they later spread throughout Asia and Europe. From early transitional types come our own more recent ancestors, some sixty thousand years ago, with developed speech, symbolic language, skills in tool-making, extended family communities along with the capacity for song and dance, and for elaborate ritual along with visual arts of amazing grandeur. All of these are expressions of the late Paleolithic Period.

Then some ten thousand years ago, the human community emerged into the Neolithic Period with its new social structures, weaving and pottery, domestication of wheat and rice, also of sheep, pigs, cattle, horses, chickens, and reindeer. Above all, this was the period of village beginnings. Out of this village context came the early cities of the world along the Tigris-Euphrates, the Nile, the Indus, the Yellow River, the Mekong. Later came the Maya on the Yucatán peninsula, the Toltec in Mesoamerica, and the Inca on the high plateaus of Peru. From its beginnings in Sumer, some five thousand years ago, the Western civilization story unfolds over the centuries, a story that leads eventually to European civilization.

The various civilizations have given expression to the human in a variety of geographical settings, with amazing inventiveness in linguistic creativity, in religious rituals, intellectual insight, social organizations, and artistic sensitivity. The Earth, during these five thousand years, has resounded with music and dance, religious spectacles, and the dramatic presentations of peoples everywhere. All this emerged as expression not simply of the human but of the Earth itself in its vast range of creativity.

Yet amid the splendor there is also the human transformation of the planet. Much of this, particularly in recent industrial centuries, has been disrupting to the functional integrity of the planet. Yet if the Earth has been exploited by cutting the forests, plowing the fields, damming the rivers, killing the animals, it has also been adorned by the pyramids of Egypt, the great temple complex of Borobudur in Indonesia, Angkor Wat in Cambodia, the Great Wall of China, the cathedrals of Europe, the Maya, Aztec, and Inca structures in the Americas. These achievements have given expression to the manner in which the peoples of the world have experienced the great mystery of things and have entered into communion with these vast cosmic forces.

Mostly these traditions have been founded in the peoples' story traditions, their mythic accounts of how things came to be in the beginning, how they came to be as they are, and how we enter with our own music, song, and dance into this unending celebration of the universe and of the planet Earth. So now in our modern scientific age, in a manner never known before, we have created our own sacred story, the epic of evolution, telling us, from empirical observation and critical analysis, how the universe came to be, the sequence of its transformations down through some billions of years, how our solar system came into being, then how the Earth took shape and brought us into existence.

With all the inadequacies of any narrative, the epic of evolution does present the story of the universe as this story is now available to us out of our present experience. This is our sacred story. It is our way of dealing with the ultimate mystery whence all things come into being. It is much more than an account of matter and its random emergence into the visible world about us. For the emergent process, as noted by the geneticist Theodore Dobzhansky, is neither random nor determined but creative. Just as in the human order, creativity is neither a rational deductive process nor the irrational wandering of the undisciplined mind but the emergence of beauty as mysterious as the blossoming of a field of daisies out of the dark Earth.

To appreciate the numinous aspect of the universe as this is communicated in this story we need to understand that we ourselves activate one of the deepest dimensions of the universe. We can recognize in ourselves our special intellectual, emotional, and imaginative capacities. That these capacities have existed as dimensions of the universe from its beginning is clear since the universe is ever integral with itself in all its manifestations throughout its vast extension in space and throughout the sequence of its transformations in times. The human is neither an addendum nor an intrusion into the universe. We are quintessentially integral with the universe.

In ourselves the universe is revealed to itself as we are revealed in the universe. Such a statement could be made about any aspect of the universe because every being in the universe articulates some special quality of the universe in its entirety. Indeed nothing in the universe could be itself apart from every other being in the universe, nor could any moment of the universe story exist apart from all the other moments in the story. Yet it is within our own being that we have our own unique experience of the universe and of the Earth in its full reality.

4

The NORTH AMERICAN CONTINENT

IN THESE OPENING YEARS OF THE TWENTY-FIRST CENTURY WE find ourselves here on this continent, known earlier as Turtle Island, now known as North America. To live here in any acceptable manner, we should know something about this continent and its distinctive features, for only in this manner can we know where we are or understand our authentic role in this context. We need to know the story of this continent, how it came to be here between the Atlantic Ocean on the east and the Pacific Ocean on the west, between the Arctic in the north and the Tropics to the south. We need to appreciate its rivers and mountains, its eastern forests, its western deserts, its Everglade swamps, its Kansas prairies, its Cascade and Sierra mountains.

Here in the east we find ourselves in the foothills of the Appalachian Mountains. Among the oldest mountains in the world, older than the Rocky Mountains or the Alps or the Himalayas, the Appalachians extend all the way down the eastern region of the continent from the Gaspé Peninsula in Canada through the long sequence of highland regions that include the White and the Green mountains in Vermont, on through the Berkshires in Massachusetts, across the Hudson River into the Catskills in New York, then down the Allegheny plateau in Pennsylvania, into the Blue Ridge Mountains of West Virginia and Carolinas into Georgia, to end in northern Alabama.

East of the mountains the coastal plain sweeps down along the Atlantic shores through New Jersey, Maryland, Virginia, the Carolinas and Georgia, then on across the Gulf Coast into Texas. To the west of the Appalachians lies the Great Central Valley where the Mississippi River and its tributaries drain the continent from the southern regions of Canada and all the way from New York in the east to Montana in the west.

At midlatitude in this valley the Prairie regions with their flowering grasses roll ever-westward over an immense area. These tall grasses originally covered the region from Western Ohio through Indiana and Illinois, then across the Mississippi into Iowa, Minnesota, and Kansas. There the mixed grasses extend on into the Dakotas, Montana, Wyoming, also into Kansas, Nebraska, and Oklahoma. The short-grass region extends from there into Western Kansas, Texas, New Mexico, and Colorado. These grasslands end where the abrupt rise of the Rocky Mountains extend their jagged peaks far above the plains. Beyond the mountains lie the desert regions of the Southwest with all their hidden and fragile yet enduring forms of life.

Just inland along the Pacific shores the mountain ranges extend along the coastal region all the way from Alaska to California. To the north the boreal spruce-fir forests extend across central Canada from Alaska to the Atlantic. In eastern Canada we find the Laurentian Hills, the most ancient rock core of this continent, also designated as the Canadian Shield, some two billion years old.

Along with such a description of the continent in its present form, we need to know how this continent came to be here, what its role has been in the past, and its apparent destiny in the unfolding future. To tell this story with any richness of detail we might begin with the time when this continent in its early form, some 250 million years ago, came together with the other land masses of the planet as a single world island, Pangaea, in the midst of the world ocean. At the meeting of the continents the Appalachian Mountains experienced their final uplift. Then some 200 million years ago the various continents rifted apart.

The North American continent swung away from the other continents toward the northwest. While separating out from the bulge of what became North Africa, this continent kept its close relationship with the Eurasian continent to the east. Indeed, Greenland geologically is part of the North American continent. Since South America later drifted off from the African continent to the southwest there has been no land contact between North America and South America until recently. North America kept its contacts with the Eurasian continent until its more complete separation some 100 million years ago. That we share the pines, the oaks, the beeches, the elms, and other tree species so extensively with the European world is due to the continued close association with that continent. As North America moved westward, the Atlantic Ocean began to take on its present form.

In the geobiological story of Earth, the period from 220 million years ago until 65 million years ago, known as the Mesozoic Era, is the period of the dinosaurs. At this immensely creative time the flowering plants and trees, and also the birds came into being in their primordial forms. Yet it was only after the massive extinctions that occurred some 67 million years ago that a new evolutionary moment occurred when the trees, flowers, birds, and animals as we know them came into being. This extinction was necessary for the emergence of many of the mammals and of our own human mode of being. At this time connections with the Asian world also began as the westward movement of the North American continent brought it closer to the continent of Asia in the far northwest.

This period of separation from the Eurasian and African worlds prevented the earlier human migrations over the Earth from reaching this continent until a rather recent period. Only when the Wisconsin, the last of the Pleistocene glaciers, moving down from the north for the last ninety thousand years, lowered the waters of the sea more than three hundred feet could the first humans enter onto this continent by crossing over a land bridge, the area now known as the Bering Straits. North America shares with South America and Australia the role of being the last of the great continents of the world to experience human presence. Africa, Asia, and Europe had long since been occupied by contemporary humans who apparently had originated much earlier in Africa.

The first peoples on this continent moved down along the valleys of the Western regions of North and then South America all the way to Tierra del Fuego at the southern tip of South America. This was possible because by this time, some six million years ago, the Isthmus of Panama connecting North America with South America had come into being and the destinies of these two continents began their association. At the same time the incoming peoples from Asia moved eastward to occupy North America all the way to the Atlantic coastlands.

The historical and cultural accomplishments of the indigenous peoples of this continent are only now beginning to be appreciated and accepted into a general narrative of the human venture. The peoples who lived here first, with their unique experience of this continent, have much to teach us concerning intimate presence to this continent, how we should dwell here in some mutually enhancing relation with the land. If the original peoples living in North and South America have not previously entered our general account of the human venture, they are now recognized as having influenced the larger course of history economically and politically as well as intellectually and spiritually.

It was the gold and silver of Central and South America that lifted the economic life of Europe to a new level of activity. The vegetables of these continents—the potatoes, corn, beans, squash, tomatoes—altered the diet of the world. The discovery of quinine, cocaine, and other healing nature products by the First Peoples of the Americas was so extensive that one writer has claimed: "This cornucopia of new pharmaceutical agents became the basis for modern medicine and pharmacology" (Weatherford, p. 184).

In our appreciation of the indigenous peoples, we might also note their achievements in the creation of languages, in their spiritual intimacy with the land, and in their political competence. In their language creations, among the most sublime and most fundamental of all human achievements, we can only marvel at the linguistic diversity. Perhaps over a thousand languages were formulated in the early period, of which over five hundred survived this early contact period. Their spiritual insight into the transhuman powers functioning throughout the natural world established the religions of Native Americans as among the most impressive spiritual traditions we know.

Their imaginative powers came to expression in their arts, literature, and dance, but especially in their poetry and their ritual. Their emotional development became manifest in those qualities of human affection and their heroic cast of soul, something consistently remarked upon by the earliest settlers. In their gentility and poise of bearing, in the affection they showed to the arriving Europeans, they made a deep impression on these first strangers from abroad.

Already in the early seventeenth century with the founding of the English colonies in the Virginia region of North America, Arthur Barlow, one of the earliest explorers of the Virginia-Carolina region, was convinced that "a more kind and loving people there cannot be found in the world" (Kolodny, p. 10). One of the most touching events in the early history of Virginia was the question posed to John Smith by the chief of the Powhatan confederacy after there had been some aggressive act by the colonists: "Why do you take by force what you may have quietly by love? Why will you destroy us who supply you with food? What can you get by war . . . ? We are unarmed and willing to give you what you ask, if you come in a friendly manner and not with swords and guns, as if to make war upon an enemy" *(Jamestown*

Voyages, edited by P. L. Barbour, p. 375 sq., quoted in T. C. McLuhan, p. 66).

While such benign traits were experienced by some of the early settlers in the eastern region of North America, these were also the heroic qualities of the Indian personality, found especially among their leaders. In the east we find Pontiac the Ottawa, who negotiated extensively with both the French and British to preserve the independence of his people. In the opening years of the nineteenth century Tecumseh the Shawnee traveled extensively and spoke to many tribes east of the Mississippi to convince them that no single tribe had a right to make a separate treaty with the English because all the land belonged in common to all the tribes. Then there is Little Turtle, the Miami war chief who defeated the forces brought against him in the Battle of Mississinewa in 1791 causing the greatest number of American casualties ever suffered in a battle with American Indians. Red Jacket the Seneca spoke with President Washington and addressed the United States Senate. To the missionaries he responded that he would wait to see how the Seneca who had been converted acted. He would then decide concerning acceptance of Christianity by himself and his people. These were leaders who spoke in council with the English with a grace and command that gave evidence of the high cultural development of the peoples here and of their capacity to address the leaders of the nations on an equal and often on a superior plane of basic cultural development.

West of the Mississippi the settlers were met by such memorable leaders as Red Cloud and Crazy Horse of the Oglala Sioux, by Black Kettle and Roman Nose of the southern Cheyenne, by Cochise and Geronimo of the Chiricahua Apache, by Chief Joseph of the Nez Percé. These are the leaders who took their peoples through the difficult times of military conflict and transition to reservation status. It is a tragic and a long continuing story that endures into the present.

Yet there is a sense in which the First Peoples of this continent in the full range of their bearing and in their intimacy with the powers of the continent have achieved something that guides and instructs all those who come to live here. Throughout these centuries despite wars, cultural oppression, poverty, and alcoholism, indigenous peoples have maintained diverse communities committed to self-determination, homelands, and ancestral traditions. These qualities of mind regard the presence of the powers of the North American continent and their traditional wisdom to be an abiding source of guidance.

This presence of native peoples to the numinous powers of this continent expressed through its natural phenomena expresses an ancient spiritual identity. The Iroquois peoples communed with these powers under the name of *Orenda,* the Algonquian as *Manitou,* the Sioux as *Wakan.* Every natural phenomenon expressed these sacred powers in some manner. To be allied with these powers is primary and necessary for every significant human endeavor on this continent.

Some sense of indigenous relation with the land can be gathered from the First Peoples' ceremonial lives, for it is in the celebrations of a people and the designs on their dress that they participate most intimately in the comprehensive liturgy of the universe. This intimacy we observe especially in the vision quest of the Plains Indians. The person entering adulthood spends several days fasting in some isolated place in hopes of receiving inner powers and a vision that would be the main source of personal strength throughout life.

We also observe this intimate relationship with the universe in the Omaha ceremony carried out at the time of birth. The infant is taken out under the sky and presented to the universe and to the various natural forces with the petition that both the universe and this continent, with all their powers, will protect and guide the child toward its proper destiny (Cronyn, pp. 53, 54). In this manner the infant is bonded with the entire natural world as the source, guide, security, and fulfillment of life.

So too, as prescribed in their Chantways, the Navaho through their sand paintings depict the entire cosmos and summon its powers to restore imbalances in the individual and in communal life. The person to be cured is placed at the center of the universe as symbolized in the painting. The cosmic powers depicted are absorbed into the person as the invocations are intoned to indicate the healing that is sought.

None of these ceremonies, however, is more appropriate or more impressive than the Iroquois Thanksgiving ceremony. This ceremony can last for days as each aspect of the natural world is addressed and gratitude is expressed for the benefits that nature has bestowed upon the people, namely the sun and the winds and the rain, the land, the streams, the growing things. In each case the relation between the natural phenomena and the human community is expressed and the gratitude of the people is offered. Natural phenomena are addressed, not in some abstract manner but in their immediate physical presence. Especially impressive is the final exhortation of gratitude offered to each component of the natural world and that the people should remember as an integral community saying "Now our minds are one." The full significance of this ceremony can only be understood if we appreciate that this celebration was the binding ritual whereby the five original tribes of the Iroquois confederacy established their unity.

The moment when the Europeans arrived on the North American continent could be considered as one of the more fateful moments in history, not only of this continent but of the entire planet. As we look back on this occasion it becomes increasingly clear that it was a moment of awesome significance, not only for the indigenous peoples, but for all the various plants and animals of this continent. Every living being on this continent might have shuddered with foreboding when that first tiny sail appeared over the Atlantic horizon.

The threatening attitude shown by the incoming settlers toward this continent as a region to be exploited in both its lands and its peoples is especially clear in the early Spanish expeditions in the southern regions of North America and in South America. In these regions conquistadors such as De Soto in the southeast, Coronado in the south central regions of North America, Cortez in Mexico, and

Pizarro in Peru were all on a relentless quest for gold. Efforts were made to enslave the Indians in their gold and silver mines and in their plantation economies. This project did not succeed because the Indians could not survive in captivity.

When in the seventeenth century the Europeans came here they might have established an intimacy with this continent and all its manifestations. They might have learned from the peoples here how to establish a viable relationship with the forests and with the forest inhabitants. They might have understood the rivers and mountains in their intimate qualities. They might have seen this continent as a land to be revered and dwelt on with a light and gracious presence. Instead it was to the colonists a land to be exploited, a theme developed by Annette Kolodny in her study of America called *The Lay of the Land.*

The continent had its moments of abundance and its moments of severity. Yet through natural processes the land was kept fertile. The trees grew to impressive girth and soaring height. The streams ran clear. The air was freshened through the seasons. The human population had its needed dwelling place. Other living beings had their appropriate habitat. From their arrival the colonists were ambivalent in their understanding of the continent. To some the land was seen as hostile, as possessed of heathen spirits. For others the land needed simply to be conquered and brought under human and Christian discipline. That the indigenous peoples were more interested in living than in working bothered the missionaries considerably. That there was no tendency to "use" the land in terms of exploitation; that there was no drive toward "progress" was a decadence not to be accepted.

Neither land nor animals nor humans can prosper in such conditions. The first of the creatures to become extinct was the auk, a defenseless bird in Newfoundland. Found in unbelievable abundance throughout several centuries these birds had nourished thousands of sailors who had taken on supplies of food here, a supply that, properly managed, could have gone on nourishing them for an indefinite period of time. Yet the drive to exploit beyond what nature could sustain was such that the great auk had disappeared by the mid-nineteenth century. The passenger pigeon, the most numerous species of birds ever known, was hunted to extinction by 1915. The buffalo, which had once numbered some sixty million, were near extinction a few years after the Civil War.

Even before all this happened, this attitude toward the land was described by William Strickland in the account of his journey up the Hudson River in 1794–1795. He wrote of the settlers in this area that they seemed to have

> an utter abhorrence for the works of creation that exist on the place where he unfortunately settles himself. In the first place he drives away or destroys the more humanized Savage the rightful proprietor of the soil; in the next place he thoughtlessly, and rapaciously exterminates all living animals, that can afford profit, or maintenance to man, he then extirpates the woods that cloath and ornament the country, and that to any but himself would be of greatest value, and finally he exhausts and wears out the soil, and with the devastation he has thus committed usually meets with his own ruin; for by this time he is reduced to his original poverty; and it is then left to him only to sally forth and seek on the frontiers, a new country which he may again devour. . . . The day appears not too distant when America so lately an unbroken forest, will be worse supplied with timber than most of the old countries of Europe. [Strickland, *Journal*, pp. 146–47]

Strickland's description is a less dramatic version of the assault on the natural world depicted by Herman Melville in *Moby-Dick*, the story of Captain Ahab's relentless pursuit of the Great White Whale. The basic attitude of the European-derived colonists was that the continent was here for human use. There was little of value to be learned from this continent that the colonists did not already know. Their schooling was in the Mediterranean-European tradition. The Greek and Latin worlds gave them their culture. The biblical Christian tradition provided spiritual meaning for their lives. Their jurisprudence, especially their sense of private property as an absolute human right, came from John Locke. The natural world itself had no rights. Nor could they even imagine that their exploitation of the continent might eventually lead to a disastrous situation for the settlers themselves.

Only a few Europeans could see that the invading peoples were disturbing the basic structure and functioning of the continent. The pervasive ideal was to structure human settlements to be occupied by a gentry that would enjoy the country to its fullest. Thus they undertook the establishment, especially in the upstate area of New York, of estates under royal patronage.

The religious groups that came in consequence of persecution in the European world recreated in the new world much of the religious antagonism toward one another that existed in Europe. Yet the continent was large enough and the various religious groups sufficiently independent that they did eventually find here the freedom from persecution and the opportunity for the new life they sought. If they were exiles, they could accept whatever hardships were involved so long as they could bring their religion with them.

Both their interior attitude and the historical situation indicated that no significant religious experience would originate in their new situation. For this to happen would be a betrayal of their Bible-derived and European-developed religion. They could not understand that their inability to commune with the land would result in the devastation of the continent. Their human-spiritual formation was complete before they came. They came supposedly with the finest religion of the world, the highest intellectual, aesthetic, and moral development, the finest jurisprudence, They needed this continent simply as a political refuge and as a region to be exploited.

Later as the incoming peoples took possession of the continent new discoveries were made. New energy sources became available. Steamships were invented. Railway lines were built. By the mid-nineteenth century modern factories and manufacturing processes were functioning. The number of peoples coming from abroad was increasing. Those who came here soon discovered the wealth of this continent in the fertility of its soils, its timber, iron ore, coal, petroleum, gold, and the other metals that became available as the industrial world of the late nineteenth century began to take shape.

The settlers were in quest of land and wealth, at whatever cost to the well-being of the continent itself. This attitude of exploitation of the land and the devastation of wildlife found expression much later in the journalist Charles Krauthammer, who in an editorial essay in *Time* magazine wrote concerning the controversy over the preservation of the spotted owl: "Nature is our ward, not our master. It is to be respected and even cultivated. But it is man's world. And when man has to choose between his well-being and that of nature, nature will have to accommodate. . . . Man should accommodate only when his fate and that of nature are inextricably bound up. In whatever situation the principle is the same; protect the environment because it is man's environment" *(Time,* 17 June 1991, p. 82).

This statement brings out the two contending attitudes toward the natural world. To indigenous peoples and to those in the founding period of the classical civilizations the natural world was the manifestation of a numinous presence that gave meaning to all existence. Human societies at whatever level of cultural sophistication found their true significance by integration of human activities into the great transformation moments of the seasonal sequence and in the movement of the day from sunrise to sunset. Human societies participated in these unending transformations. They simply gave intelligent recognition of that spirit presence pervading the entire natural world. The natural world provided both the physical and the psychic needs of humans. These were inseparable gifts that came to humans in the same moment and through the same causes.

As seen by the Europeans the continent was here to serve human purposes through trade and commerce as well as through the more immediate personal and household needs of the colonists. They had nothing spiritual to learn from this continent. Their attitude toward the land as primarily for *use* was the crucial issue. This attitude was not only the clash of two human groups with each other over some land possession or some political rule, it was a clash between the most anthropocentric culture that history has ever known with one of the most naturecentric cultures ever known. It was the clash between a monotheistic personal deity perceived as transcendent to all phenomenal modes of being and the Great Spirit perceived as immanent in all natural phenomena. It was the clash between a people driven by a sense of historical destiny with a people living in an abiding world of ever-renewing phenomena. It was the clash of a people with certain immunities to tuberculosis, diphtheria, and measles with a people devoid of such immunities. Over the centuries it became the clash of an urban people highly skilled in industrial manufacturing with a tribal people skilled in hunting and farming who could still appreciate the integral relations that exist between the human community and the natural world.

The insuperable difficulty inhibiting any intimate rapport with this continent or its people was this European-derived anthropocentrism. Such orientation of Western consciousness had its fourfold origin in the Greek cultural tradition, the biblical-Christian religious tradition, the English political-legal tradition, and the economic tradition associated with the new vigor of the merchant class. In religion, culture, politics, and economics there existed with the settlers a discontinuity of the human with the natural world. The human, transcendent to the natural world, was the assumed ruler of the land.

That is why the North American continent became completely vulnerable to the assault from the European peoples. To the European settlers the continent had no sacred dimension. It had no inherent rights. It had no way of escaping economic exploitation. The other component members of the continent could not be included with humans in an integral continental community. European presence was less occupation than predation.

The critique of this attitude came through the naturalist writers, the poets and artists, and on occasion in some of the writings and sermons of ministers such as Jonathan Edwards (1703–1758), to whom all creation manifested the divine glory. Yet these critiques were peripheral to the basic orientation of American thought and culture throughout this period. Even the transcendentalist essayists were less inspired by the continent than is sometimes thought. In the four areas of life enumerated (religion, culture, politics, and economics), the cultural commitments are so deep in the American soul, so imprinted in the unconscious depths of the culture, that until now it has not been possible to critique these areas of human endeavor in any effective manner. We saw ourselves as the envy of the ages, as relieved of superstition and in the highest realms of intellectual enlightenment.

So committed were we to our divinely commissioned task of commercially exploiting this continent that we could even experience a high spiritual exaltation in what we were doing. Even now, becoming an entrepreneur is sometimes experienced as a religious "rite of passage." When one successful businessman who had previously been trained in several different spiritual disciplines was asked, "Which of your experiences gave you the biggest spiritual charge?" His answer was, "Entrepreneurship. When I became an entrepreneur it became clear that everything I had done before that had just been *NATO*." On further inquiry concerning the meaning of that expression, he answered, "No action—talk only. When I put myself on the line totally as an entrepreneur committing my money and energy and time to the vision I had—I saw the falseness of ever doing less than that" *(The Tarrytown Letter,* May 1984, p. 14).

Yet after these past few centuries of European presence here, we begin to rethink our situation. We look about us and see a continent severely devastated in its primordial forests, its air polluted with residue from fossil fuel–based power systems, its soils impoverished by chemical fertilizers, its rivers dammed for irrigation and for their hydroelectric energy, polluted by runoff from the fertilizers, herbicides, and pesticides used in agriculture, and poisoned by the mercury used in panning for gold. In March of 1999 nine types of salmon were listed under the Environmental Protection Act as threatened with extinction due to pollution and runoff that includes sewage and chemicals from fertilizers and by construction of dams that simply closed off ancient rivers leading to the breeding grounds of the salmon. Then there are the factory ships that scoop up enormous quantities of salmon on their return journey to the rivers of their origin. Immense shoals of fish in the North Atlantic are also disappearing. In its survey of the global situation as regarding the fisheries of the globe, World Resources Institute in its publication *World Resources 1998–1999* tells us "World fisheries face a grim forecast. Forty-five years of increasing fishing pressure have left many major fish stocks depleted or in decline" (p. 195). "The harvest of overexploited fish stocks has dropped 40% in only 9 years." Also noted in that same source is the fact that the Atlantic cod, haddock, and bluefin tuna were listed by the World Conservation Union in 1966 "as species whose survival is to some extent endangered" (p. 196).

This situation with the fishing industry is simply one instance of a more encompassing issue that we must deal with in these emerging years of the twenty-first century. Yet a pervasive awareness of the damage we have done begins to sweep over the land. We have taken possession of the continent but find ourselves becoming deprived of those luxuriant life sources that once surrounded us. Prairie lands with fertile soils deep into the Earth have lost a third of the soils they once had. The streams that once flowed fresh and pure and life-giving are now polluted and poisonous to all forms of life. The aquifers beneath the western plains are being drained to exhaustion.

We might well brood over these scenes until we come to some depth awareness of what has happened and begin to dream again, this time a more coherent dream of an integral community of the human and all those other-than-human component members of the North American continent.

5

The WILD *and the* SACRED

TO UNDERSTAND THE HUMAN ROLE IN THE FUNCTIONING OF the Earth we need to appreciate the spontaneities found in every form of existence in the natural world, spontaneities that we associate with the wild—that which is uncontrolled by human dominance. We misconceive our role if we consider that our historical mission is to "civilize" or to "domesticate" the planet, as though wildness is something destructive rather than the ultimate creative modality of any form of earthly being. We are not here to control. We are here to become integral with the larger Earth community. The community itself and each of its members has ultimately a wild component, a creative spontaneity that is its deepest reality, its most profound mystery.

We might reflect on this sense of the wild and the civilized when the dawn appears through the morning mist. At such times a stillness pervades the world—a brooding sense, a quiet transition from night into day. This experience is deepened when evening responds to morning, as day fades away, and night comes in the depth of its mystery. We are most aware at such moments of transition that the world about us is beyond human control. So too in the transition phases of human life; at birth, maturity, and death we brood over our presence in a world of mystery far greater than ourselves.

I bring all this to mind because we are discovering our human role in a different order of magnitude. We are experiencing a disintegration of the life systems of the planet just when the Earth in the diversity and resplendence of its self-expression had attained a unique grandeur. This moment deserves special attention on the part of humans who are themselves bringing about this disintegration in a manner that has never happened previously in the entire 4.6 billion years of Earth history.

We never thought of ourselves as capable of doing harm to the very structure of the planet Earth or of extinguishing the living forms that give to the planet its unique grandeur. In our efforts to reduce the planet to human control we are, in reality, terminating the Cenozoic Era, the lyric period of life development on the Earth.

If such moments as dawn and dusk, birth and death, and the seasons of the year are such significant moments, how awesome, then, must be the present moment when we witness the dying of the Earth in its Cenozoic expression and the life renewal of the Earth in an emerging Ecozoic Era. Such reflection has a special urgency if we are ever to renew our sense of the sacred in any sphere of human activity. For we will recover our sense of wonder and our sense of the sacred only if we appreciate the universe beyond ourselves as a revelatory experience of that numinous presence whence all things come into being. Indeed, the universe is the primary sacred reality. We become sacred by our participation in this more sublime dimension of the world about us.

The universe carries in itself the norm of authenticity of every spiritual as well as every physical activity within it. The spiritual and the physical are two dimensions of the single reality that is the universe itself. There is an ultimate wildness in all this, for the universe, as existence itself, is a terrifying as well as a benign mode of being. If it grants us amazing powers over much of its functioning we must always remember that any arrogance on our part will ultimately be called to account. The beginning of wisdom in any human activity is a certain reverence before the primordial mystery of existence, for the world about us is a fearsome mode of being. We do not judge the universe. The universe is even now judging us. This judgment we experience in what we refer to as the "wild." We recognize this presence when we are alone in the forest, especially in the dark of night, or when we are at sea in a small craft out of sight of land and for a moment lose our sense of direction. The wild is experienced in the earthquakes that shake the continents in such violence, so too in the hurricanes that rise up out of the Caribbean Sea and sweep over the land.

We have at times thought that we could domesticate the world, for it sometimes appears possible, as in our capacity to evoke the vast energies hidden in the nucleus of the tiny atom. Yet when we invade this deepest, most mysterious dimension of matter, nature throws at us its most deadly forces, wild forces that we cannot deal with, forces that cause us to fear lest we be rendering the planet a barren place for the vast range of living beings.

I speak of the wild dimension of existence and the reverence and fear associated with the wild, since precisely here is where life and existence and art itself begin. When Thoreau in his essay on walking said, "In Wildness is the Preservation of the World," he made a statement of unsurpassed significance in human affairs. I know of no more comprehensive critique of civilization, this immense effort that has been made over these past ten thousand years to bring the natural world under human control. Such an effort would even tame the inner wildness of the human itself. It would end by reducing those vast creative possibilities of the human to trivial modes of expression.

Wildness we might consider as the root of the authentic spontaneities of any being. It is that wellspring of creativity whence come the instinctive activities that enable all living beings to obtain their food, to find shelter, to bring forth their young; to sing and dance and fly through the air and swim through the depths of the sea. This is the same inner tendency that evokes the insight of the poet, the skill of the artist, and the power of the shaman. Something in the wild depths of the human soul finds its fulfillment in the experience of nature's violent moments.

As one woman told a group assembled in Florida after Hurricane Andrew, she did not consider herself a victim but a participant in this wild event in all its creative as well as its destructive aspects. The hurricane, she insisted, was telling us something. It was telling us how to build our houses if we wished to dwell in this region. It was telling us to consider well the winds and the sea, to mark well the fact that if we live here we must obey the deeper laws of the place, laws that cannot be overridden by any type of human zoning. We might live here if we wish but on terms dictated by powers other than human. The hurricane has its own inner discipline. It is itself a response to the needs of the region. This we need to know: how to participate creatively in the wildness of the world about us. For it is out of the wild depths of the universe and of our own being that the greater visions must come.

We mistake the wild if we think of it as mere random activity or simply as turbulence. Throughout the entire world there exists a discipline that holds the energies of the universe in the creative pattern of their activities, although this discipline may not be immediately evident to human perception. The emergent universe appears as some wild, senseless deed that wells up from some infinite abyss in the expansive differentiating process of those first moments when all the energy that would ever exist flared forth in a radiation too mysterious for humans to fully comprehend. Yet as this energy articulated itself in the form of matter the gravitational attraction that each being has for every other being produced the basic ordering process, gravitation, the primary discipline in the large-scale structure of the universe.

This mutual attraction and mutual limitation of gravitation is, perhaps, the first expression and the primordial model of artistic discipline. It gave to the universe its initial sense of being at home with itself and yet caught up in a profound discontent with any final expression of itself. We might consider, then, that the wild and the disciplined are the two constituent forces of the universe, the expansive force and the containing force bound into a single universe and expressed in every being in the universe.

This too is the ultimate secret of the planet Earth. When first the solar system gathered itself together with the sun as the center surrounded by the nine fragments of matter shaped into planets, the planets that we observe in the sky each night, these were all composed of the same matter; yet Mars turned into rock so firm that nothing fluid can exist there, and Jupiter remained a fiery mass of gases so that nothing firm can exist there.

Only Earth became a living planet filled with those innumerable forms of geological structure and biological expression that we observe throughout the natural world. Only Earth held a creative balance between the turbulence and the discipline that are necessary for creativity. The excess of discipline suppressed the wildness of Mars. The excess of wildness overcame the discipline of Jupiter. Their creativity was lost by an excess of one over the other. Yet an equilibrium of these forces would have brought about another barrier to creativity, for equilibrium would produce a fixation in which creativity would be lost. The universe solved its problem by establishing a creative disequilibrium expressed in the curvature of space that was sufficiently closed to establish an abiding order in the universe and sufficiently open to enable the wild creative process to continue.

We perceive this creative power primarily in the intelligible order we observe in the universe. Such is the way of the philosopher. Such is the way of Saint John in the opening prologue to his gospel, "In the beginning was the *Word*," the principle of order and intelligibility. Or we can perceive the originating power itself in the disequilibrium of the universe, in the spirit world, in the wildness of things, in the dreams that come into our souls in the depths of the night, dreams that correspond in the human soul to the openness of the curve that holds the universe together and yet enables it to continue its infinite creativity.

Artists have something in them that is wild, something guided and inspired ultimately by imagination. "Divine imagination" in the words of William Blake. The artist revels in the ultimate disequilibrium of things. The philosopher is controlled ultimately by the balance and harmony of things, by reasoning intelligence. Both are valid, both are needed. The universe from the beginning, and even now, is poised between the expanding and the containing forces, and no one knows just when or if this creative balance will collapse or will continue on indefinitely. So the philosopher and the artist are both poised between the two possibilities.

In this mysterious balance the universe and all its grandeur and all its loveliness become possible. Exactly here the presence of the sacred reveals itself. Here is the exuberance that could fling the stars across the heavens with such abandon and yet with such exquisite poise, each in relation to the untold billions of other shining fragments of primordial existence. Musicians who listen to the wild rhythms and melodies that arise within them experience the power forces of the universe. Such we find with Johann Sebastian Bach in his Toccata and Fugue in D Minor, with Ludwig van Beethoven in his *Eroica* Symphony; with Van Gogh in his *Starry Night*. Such too is the magic of Claude Monet in his evocation of the mystical qualities of waterlilies floating on a small dark pond. The architect who looked out over the fields south of Paris and saw, through the mist of the future, the Chartres Cathedral must have experienced a moment when the wild and the sacred came together in a single moment of vision.

A similar experience of the wild wonders of the universe can be seem on a smaller scale in the dream paintings of the Aborigines of

Australia. Here in the desert regions of this vast continent in the southern seas are a people who experience the universe around them, especially the topography of the land, as expressions of those preternatural beings or powers referred to as Dreamings. Their paintings, composed of lines and dots in an endless variety of patterns in color and design, are unimaginable in our Western traditions. These paintings portray the Dreamings as the creative forces producing the landscape and expressive of the deepest spirit of the universe.

The indigenous peoples of Australia were once thought to be totally lacking intellectual or cultural achievements associated with even the most primitive peoples known elsewhere. They had food only for the day, no permanent dwellings, no clothes, only a few implements. Yet we now find amazing achievements in their capacity for understanding and responding to both the physical and the spiritual dynamics of the world about them. Most significant of all is the fact that they live in a universe.

To emphasize this might seem needless, since everyone lives in the universe; but seldom do we have any real sense of living in a world of sunshine by day and under the stars at night. Seldom do we listen to the wind or feel the refreshing rain except as inconveniences to escape from as quickly as possible. In the city, snow in all its purity quickly becomes a dirty thing with all the drifting particles of human-caused exhaust and residue that settle upon it.

The world of mechanism has alienated us from the wild beauty of the world about us. Such is the power of art, however, that it can endow even the trivialities and the mechanisms of our world with a pseudomystical fascination. Supposedly this enables our world to avoid the epithet of being caught in an imitative Classicism or in a faded Romanticism. The result is to challenge any traditional norms of beauty by a reversion to the wild simply through undisciplined turbulence, at times with an elaborate presentation of the trivial, or even with what is referred to as a personal statement.

The landscape that encloses the Appalachian region, the rivers that flow down from the mountains to the sea, the trees that blossom in these surroundings, the birds that sing throughout this valley, all these were brought into being during this past 65 million years. If this has been a period of wildness beyond compare, it has also been the lyric period in the story of Earth. The human, perhaps, could only have appeared in such a period of grandeur; for the inner life of the human depends immediately on the outer world of nature. Only if the human imagination is activated by the flight of the great soaring birds in the heavens, by the blossoming flowers of Earth, by the sight of the sea, by the lightning and thunder of the great storms that break through the heat of summer, only then will the deep inner experiences be evoked within the human soul.

All these phenomena of the natural world fling forth to the human a challenge to be responded to in literature, in architecture, ritual, and art, in music and dance and poetry. The natural world demands a response beyond that of rational calculation, beyond philosophical reasoning, beyond scientific insight. The natural world demands a response that rises from the wild unconscious depths of the human soul. A response that artists seek to provide in color and music and movement.

The response that we give must have a supreme creative power, for the Cenozoic Era in the story of Earth is fading as the sun sets in the western sky. Our hope for the future is for a new dawn, an Ecozoic Era, when humans will be present to the Earth in a mutually enhancing manner.

6

The VIABLE HUMAN

WE NEED TO MOVE FROM OUR HUMAN-CENTERED TO AN EARTH-centered norm of reality and value. Only in this way can we fulfill our human role within the functioning of the planet we live on. Earth, within the solar system, is the immediate context of our existence. Beyond the sun is our own galaxy and beyond that the universe of galactic systems that emerged into being some fifteen billion years ago through some originating source beyond human comprehension.

Establishing this comprehensive context of our thinking is important in any consideration of human affairs, for only in this way can we identify any satisfying referent in our quest for a viable presence of the human within the larger dynamics of the universe. The universe itself is *the* enduring reality and *the* enduring value even while it finds expression in a continuing sequence of transformations.

By bringing forth the planet Earth, its living forms, and its human intelligence, the universe has found, so far as we know, its most elaborate expression and manifestation of its deepest mystery. Here, in its human mode, the universe reflects on and celebrates itself in a unique mode of conscious self-awareness. Our earliest documents reveal a special sensitivity in human intellectual, emotional, and aesthetic responses to this larger context of survival. The Universe, the Earth, the Human are centered in one another. The later realms of being are dependent on the earlier for survival while the earlier realms are dependent on the later for their more elaborate manifestation. The more complex are dependent on the more simple; the more simple are revealed in the more elaborate.

Instinctively, humans perceived themselves as a mode of being of the universe as well as distinctive beings in the universe. This was the beginning. The emergence of the human was a transformation moment for the Earth as well as for the human. As with every species, there was a need for humans to establish their niche, a sustainable position in the larger community of life, a way of obtaining food, shelter, and clothing. There was need for security, the need for family and community context. This need for community was quite special in the case of humans since humans articulate a capacity for thought and speech, aesthetic appreciation, emotional sensitivities, and moral judgment, none of which can function without a community context. If we speak there must be someone to teach us, someone to listen. To sing or make music is a personal fulfillment, but it is most satisfying if we are sharing with others or communicating some deep feeling to others. There is only community thinking and poetry and dancing, a community with historical continuity over a series of generations. These combine in a cultural shaping that establishes the human with identifying qualities, a shaping that is communicated to succeeding generations by family nurturing and by formal teaching.

Whatever the cultural elaboration of the human, basic physical as well as psychic nourishment and support come from the natural environment. When we speak of the natural world we are not speaking simply of the physical world but of the psychic-physical mode of being found in every articulated entity of the phenomenal world. Human society in its beginnings would not have survived if it had not had some basic role to fulfill within the larger Earth community composed of all its geological as well as its biological elements. If from its earliest period the human put a certain stress upon other forms of life, this is quite coherent with the order of things, since such stress occurs within the general norms of interrelation of species.

Once we recognize that a change from a human-centered to an Earth-centered norm of reality and value is needed, we might ask how this is to be achieved and how it would function. We might begin by recognizing that the life community, the community of all living species, including the human, is the greater reality and the greater value. The primary concern of the human community must be the preservation and enhancement of this comprehensive community, even for the sake of its own survival.

While humans do have their own distinctive reality and unique value, these must be articulated within a more comprehensive context. Ultimately humans find their own being within this community context. To consider that one is enhanced by diminishing the other is an illusion. Indeed, it is the great illusion of the present industrial age, which seeks to advance human well-being by plundering the planet in its geological and biological structure and functioning.

Opposed to this exploitation of the natural world is the ecological movement, which seeks to create a more viable context for human development within the planetary process. We must clearly understand, however, that this question of viability is not an issue that can be resolved in any permanent manner. It will be a continuing issue for the indefinite future. Indeed, we are at the present time participating in an unparalleled change in the human-Earth situation. The planet that ruled itself directly over these past millennia is now determining its future largely through human decision. Such is the responsibility assumed by the human community once we ventured onto the path of the empirical sciences and their associated technologies. In this process, whatever the benefits, we endangered ourselves and every living being on the planet. We altered the entire mode of functioning of the planet.

If we look back over the total course of Earth development, we find that there was a consistent florescence of the life process in the larger arc of the planet's development over some billions of years. Innumerable catastrophic events occurred in both the geological and biological realms but none of these could cause the forebodings such as we might experience at present. There is no question of the extinction of life in any total sense, even though many of the more elaborate forms of life expression can be eliminated in a permanent manner. What is absolutely threatened just now is the degradation of the planet. This degradation involves extensive distortion and a pervasive weakening throughout the entire life system of the planet.

Because such deterioration results from a rejection of the inherent limitations of human existence and from an effort to alter the natural functioning of the planet in favor of a humanly constructed wonderworld, resistance to this destructive process must turn its efforts toward living creatively within the organic functioning of the natural world. Earth as a biospiritual planet must become for us the basic referent in identifying our own future.

At present we have the ecologist and a large number of ecological organizations standing over against the industrial, commercial, and financial corporations in defense of a viable mode of human functioning within the planetary process. This opposition between the industrial-commercial entrepreneur and the ecologist can be considered as both the central human issue and the central Earth issue of the twenty-first century. It seems quite clear that after these centuries of industrial efforts to create a wonderworld we are in fact creating wasteworld, a nonviable situation for the human mode of being. The true wonderworld of nature, whatever its own afflictions, is available as the context for a viable human situation. The difficulty just now is that the financial and industrial establishments have such extensive control over the planet that change so basic as that suggested here would be extremely difficult.

After identifying the order of magnitude of the difficulty before us, we need to establish a more specific analysis of the problems themselves. Then we need to provide specific programs leading toward a viable human situation on a viable planet. For this purpose I offer the following analysis of the present situation as it exists under the controlling power of the industrial entrepreneur and then offer alternative proposals for a viable human situation.

As concerns *natural resources,* the industrial, commercial, and financial corporations are in possession of the planet; either directly or indirectly, with the support of governments subservient to the various corporation enterprises. This possession is, of course, within limits. Fragmentary regions of the planet have been set aside or will be set aside as areas to be preserved in their natural state or to be exploited at a later time. Yet these regions themselves, frequently enough, survive by consent of the controlling corporations.

To the ecologist, reducing the planet to a resource base for consumer use in an industrial society is already an unacceptable situation. The planet and all its components are reduced to commodities whose very purpose of existence is to be exploited by the human. Our more human experience of the world of meaning has been diminished in direct proportion as money and utilitarian values have taken precedence over the numinous, aesthetic, and emotional values. In a corresponding way, any recovery of the natural world in its full splendor will require not only a new economic system but a conversion experience deep in the psychic structure of the human.

Our present situation is the consequence of a cultural fixation, an addiction, an emotional insensitivity, none of which can be remedied by any quickly contrived adjustment. Nature has been severely, and in many cases irreversibly, damaged. A healing is often available and new life can sometimes be evoked, but this cannot be without an intensity of concern and sustained vigor of action such as that which brought about the damage in the first place. Without this healing, the viability of humans at any acceptable level of fulfillment remains in question.

As regards *law,* the basic orientation of American jurisprudence is toward personal human rights and toward the natural world as existing for human possession and use. To the industrial-commercial world the natural world has no inherent rights to existence, habitat, or freedom to fulfill its role in the vast community of existence. Yet there can be no sustainable future, even for the modern industrial world, unless these inherent rights of the natural world are recognized as having legal status. The entire question of possession and use of the Earth, either by individuals or by establishments, needs to be considered in a more profound manner than Western society has ever done previously.

The naive assumption that the natural world is there to be possessed and used by humans for their advantage and in an unlimited manner cannot be accepted. The Earth belongs to itself and to all the component members of the Earth community. The Earth is there as an entrancing celebration of existence in all its alluring qualities. Each earthly being participates in this cosmic celebration as the proper fulfillment of its powers of expression. Reduction of the Earth to an object primarily for human possession and use is unthinkable in most traditional cultures. Yet to Peter Drucker, author of *Innovation and Entrepreneurship,* the entrepreneur creates resources and values. Before it is possessed and used, "every plant is a weed and every mineral is just another rock" (Drucker, p. 38). In this context, human possession and use is what activates the true nobility of any natural object.

To achieve a viable human-Earth situation a new jurisprudence must envisage its primary task as that of articulating the conditions for the integral functioning of the Earth process, with special reference to a mutually enhancing human-Earth relationship. Within this context the various components of the Earth—the land, the water, the air, and the complex of life systems—would each be a commons. Together they would constitute the integral expression of the Great Commons of the planet Earth to be shared in proportion to need among all members of the Earth community.

In this context each individual being is supported by every other being in the Earth community. In turn, each being contributes to the well-being of every other being in the community. Justice would consist in carrying out this complex of creative relationships. Within the human community there would be a need for articulating patterns of social relationships in which individual and group rights would be recognized and defended. The basic elements of personal security and personal property would be protected, although the sense of ownership would be a limited personal relation to property, which would demand use according to the well-being of the property and the well-being of the community, along with the well-being of the individual owner. The entire complex of political and social institutions would be needed. Economic organizations would also be needed. But these would all be so integral with the larger Earth economy that they would enhance rather than obstruct one another.

Another significant aspect of contemporary life, wherein the entrepreneur has a dominant position, is in *language*. Since we are enclosed in an industrial culture and a consumer economy, the words we use have their significance and validation within this culture. A central value word used by our society is that of *progress*. This word has a wide range of significance as regards our increasing scientific understanding of the universe, our increase in personal and social development, our attainment of better health and longer life. Through our modern technologies, we can manufacture greater quantities of products with greater facility. We can travel faster and with greater ease. So we continue to progress endlessly with a feeling that all is well.

But then we see that our human progress has been carried out by desolating the natural world. This "degradation" of the Earth is seen as the condition for "progress" of humans. In his opening statement giving an overview of the world situation Lester Brown in *Vital Signs* tells us, "The world today is warmer, more crowded, more urban, economically richer and environmentally poorer than ever before" (Brown, p. 15). The Earth is a kind of sacrificial offering. Within the human community, however, there is little awareness that the integral survival of the planet in its seasonable rhythms of renewal is itself a condition not simply of human progress but of human survival. Often the ecologist is at a loss as to how to proceed; the language in which our values are expressed has been co-opted by the industrial establishment and is used with the most extravagant modes of commercial advertising to create the illusory world in which modern industrial peoples now live.

One of the most essential roles of the ecologist is to create the language in which a true sense of reality, of value, and of progress can be communicated to our society. This need for rectification of language in relation to reality was recognized early by the Chinese as the first task of any acceptable guidance for the society *(Analects* XXII: 11). Just now, a rectification is needed in the term *progress*. There is a sense in which progress is needed in relieving humans from some of the age-old afflictions that humans have borne. Yet this sense of progress is being used as an excuse for imposing awesome destruction on the planet for the purpose of monetary profit, even when the consequences involve new types of human psychic and physical misery.

The term *profit* needs to be rectified. Profit according to what norms and for whom? The profit of the corporation is the deficit of the Earth. The profit of the industrial enterprise, whatever its advantages, can also be considered as a deficit in the quality of life. We need to reexamine the entire range of our language.

There are questions concerning "gender" that need consideration. The industrial establishment is the extreme expression of a patriarchal tradition with its all-pervasive sense of dominance, whether of rulers over people, of men over women, of humans over nature. Only with enormous psychic and social effort and revolutionary processes has this patriarchal control been mitigated as regards the rights of women. The rights of the natural world of living beings is still at the mercy of the modern industrial corporation as the ultimate expression of patriarchal dominance over the entire planetary process.

Then too we begin to recognize the rights of ethnic groups and of the impoverished classes of our society. For the ecologist, the great model of all existence is the natural ecosystem, which is self-ruled as a community in which each component has its unique rights and its comprehensive influence. The ecologist, with a greater sense of the human as a nurturing presence within the larger community of the geological and biological modes of Earth being, is sponsoring a mode of human activity much closer to the feminine than to the masculine modes of being and of activity.

As concerns *education,* its purpose as presently envisaged, is to enable persons to be "productive" within the context of the industrial society. A person needs to become literate in order to fulfill some function within the system, whether in acquisition or processing of raw materials, manufacturing, distributing the product in a commercially profitable manner, managing the process or the finances, or, finally, spending the net earnings in acquisition and enjoyment of possessions. A total life process is envisaged within the industrial process. All professional careers now function within the industrial-commercial establishment, even education, medicine, and law.

In this new context of a viable human mode of being, the primary educator as well as the primary lawgiver and the primary healer would be the natural world itself. The integral Earth community would be a self-educating community within the context of a self-educating universe. Education at the human level would be the sensitizing of the human to those profound communications made by the universe about us, by the sun and moon and stars, the clouds and rain, the contours of the Earth and all its living forms.

The music and poetry of the universe would flow into the student; also a sense of the deep mystery of existence as well as insight into the architecture of the continents and the engineering skills whereby the great hydrological cycles function in moderating the temperature of the Earth, in providing habitat, and in nourishing the multitudes of living creatures.

This orientation toward the natural world should be understood in relation to all human activities. The Earth would be our primary teacher in industry and economics. It would teach us a system in which we would create a minimum of entropy, a system in which there would be a minimum of unusable or unfruitful waste. Such is the system of waste treatment evolved by John and Nancy Todd in their design of "living machines" that recycle waste water. This is done by means of artificial wetlands where plants take the toxic elements from the waste. Already such wetlands are functioning in several New England cities.

In this context we need to appreciate also the work of Miriam Therese MacGillis at Genesis Farm. This involved a Community Supported Agriculture project that has established an integral relation of the local community with the region it occupies and with the beginnings of a subsistence economy. We might also mention the urban gardening project of Paul and Julie Mankiewicz in New York, the large-scale organic agricultural project of Fred Kirschenmann in the northern Great Plains regions, the teaching and examples of renewable energy projects of Amory and Hunter Lovins at the Rocky Mountain Institute in the Snowmass area of Colorado. Amory Lovins's *Soft Energy Paths* is something of a classic reference work. By the example of his farm and the teachings in his book, *The Unsettling of America: Culture and Agriculture,* Wendell Berry provides a superb illustration of what can be done by settling on the land in some integral manner.

In the realm of city architecture Richard Register has done remarkable work in his sponsoring of four international conferences on this subject. His *Eco-City Berkeley* provides some indication of the possibilities that are available. Others, such as Peter Berg in his writings, have provided the initiative and guidance for the renewal of urban life. One of the best modern examples of a metropolitan city that has recovered its integrity as a community into the twenty-first century is Curitiba, a city of some two million people in southern Brazil, inland seventy miles from the coast. By recovering the inner city as a living place where celebration could provide the delight in life that is needed by any community, Curitiba has since developed into a viable modern city. Most significant is the sense of a shared way of life that identifies Curitiba. Public expense per person is estimated at one-tenth of the expense per person in Detroit.

Most impressive in this general context is the small country of Ladakh. Helena Norberg-Hodge, one of the few outsiders who knows the language well enough to enter into their way of life in some intimate manner, has written a remarkable account of these people in *Ancient Futures: Learning from Ladakh.* Ladakh lies east of Kashmir high in the Himalayas on the border of China, and has a Tibetan Buddhist religious orientation. It has a severe climate of eight months of winter with temperatures of 40 degrees below freezing, little rainfall, and a meager soil—all indicating as harsh a natural condition as a person might imagine. Yet the people of Ladakh have educated themselves in a sustainable way of life, a grace and delight in living, a sense of community, and thus have created a world of meaning in its deepest sense. All this provides convincing evidence that education in direct relation with the natural environment and the use of basic technologies can supply the needs of life in a context that leads to a high level of personal fulfillment. Success or failure as human communities has no absolute need of such technologies or conveniences as we imagine are needed for a proper life fulfillment.

Much more could be said concerning our relations with the surrounding environment and the need to live integrally with the natural world. But this may be sufficient to suggest a context for thinking about the surrounding world within an ecologically sound system of education, an education that would be available to everyone from the beginning to the end of life when the Earth that brought us into being draws us back into itself to enter the deepest of all mysteries.

In our view of a viable future, a new context would exist also for *the medical profession.* The problems of human illness are not only increasing but are being altered in their very nature by the industrial context of life. In prior centuries, human illness was experienced within the well-being of the natural world with its abundance of air and water, and foods grown in fertile soil. Even city dwellers with their deteriorated natural surroundings could depend on the purifying processes of the natural elements. The polluting materials themselves were subject to natural decomposition and reabsorption into the ever-renewing cycles of the life process. But this is no longer true. The purifying processes have been overwhelmed by the volume, the composition, and the universal extent of the toxic or nonbiodegradable materials. Beyond all this, the biorhythms of the natural world are suppressed by the imposition of mechanistic patterns on natural processes.

The profession of medicine must now consider its role, not only within the context of human society, but in the context of the Earth process. A healing of the Earth is a prerequisite for the healing of the human. Adjustment of the human to the conditions and restraints of the natural world constitutes the primary medical prescription for human well-being. The medical profession needs to establish a way of sustaining the species as well as the individual if the human is to be viable as a species within the community of species.

Behind the long disruption of the Earth process is the refusal of Western industrial society to accept needed restraints upon its quest for release, not simply from the normal ills to which we are subject, but release from the human condition itself. There exists in our tradition a hidden rage against those inner as well as outer forces that create limits on our activities.

Some ancient force in the Western psyche seems to perceive limitation as a demonic obstacle to be eliminated, rather than as a strengthening discipline. Acceptance of the challenging aspect of the natural world is a primary condition for creative intimacy with the natural world. Without this opaque or even threatening aspect of the universe we would lose our greatest source of creative energy. This opposing element is as necessary for us as is the weight of the atmosphere that surrounds us. This containing element, even the gravitation that binds us to the Earth, should be experienced as liberating and energizing, rather than confining.

Strangely enough, it is our efforts to establish a thoroughly sanitized world that have led to our toxic world. Our quest for wonderworld is making wasteworld. Our quest for energy is creating entropy on a scale never before witnessed in the historical process. We have invented a counterproductive society that is now caught in the loop that feeds back into itself in what can presently be considered a runaway situation.

The media and advertising are particularly responsible for placing the entire life process of the human in a situation wherein producer and consumer feed back into each other in an ever-accelerated process. Presently we experience on a world scale an enormous glut in many basic products, along with unmatched deprivation in the vast numbers of peoples gathered in the shantytowns of the world.

Few of our most prominent newspapers, newsweeklies, or periodicals written for the general public have a consistently designated space for the ecological situation, although environmental concerns are being mentioned more frequently. While there are regular sections for politics, economics, sports, arts, science, education, food, entertainment, and a number of other areas of life, only on rare occasions do significant articles appear concerned with what is happening to the planet.

These periodicals are of course supported by the great industrial establishments. Media attention to the disturbed life systems of the Earth is considered as threatening or limiting to the industrial enterprise. In this situation the commercial-industrial control of the media can be considered among the most effective forces thwarting any remedial action to save the disintegrating planet.

There are efforts to mitigate the problematic consequences of this industrial-commercial process by modifying the manner in which these establishments function, efforts such as the reduction in the amount of toxic waste produced, as well as more efficient modes of storing or detoxifying waste. Yet all of this is trivial in relation to the magnitude of the issues. So too are the regulatory efforts of governments. These have consistently been microphase solutions for macrophase problems.

We witness also the pathos of present efforts at preserving habitat for wildlife in some countries while the tropical rain forests are being destroyed each year in other countries. Among the other efforts at altering our present destructive activities are the confrontational groups such as Greenpeace, Sea Shepherd, and Earth First!. These are daring ventures that dramatize the situation in its stark reality. That such tactics are needed to force a deeper reflection on what we are doing is itself evidence of how profound a change is needed in human consciousness.

Beyond such mitigating efforts and such confrontational tactics is the clarification of more creative modes of functioning in all our institutions and professions. This clarification is already taking place through movements such as those concerned with reinhabiting the various bioregions of the world. All of these new and mutually enhancing patterns of human-Earth relationships are being developed on a functional as well as a critical-intellectual basis. This orientation begins to find expression in politics, in economics, in education, in healing, and in spiritual reorientation. Together these movements, oriented toward a more benign human relationship with the environment, indicate a pervasive change in consciousness that presently is our best hope for developing a sustainable future.

Meanwhile, in the obscure regions of the unconscious where the primordial archetypal symbols function as ultimate controlling factors in human thought, emotion, and in practical decision-making, a profound reorientation toward this integral human-Earth relationship is gradually taking place. The archetypal journey of the universe can now be experienced as the journey of each individual, since the entire universe has been involved in shaping our individual psychic as well as our physical being from that first awesome moment when the universe emerged.

We might now recover our sense of the maternal aspect of the universe in the symbol of the Great Mother, especially in the Earth as that maternal principle out of which we are born and by which we are sustained. Once this symbol is recovered the dominion of the patriarchal principle that has brought such aggressive attitudes into our activities will be mitigated. If this is achieved then our relationship with the natural world would undergo one of its most radical readjustments since the origins of our civilization in classical antiquity.

We might also recover our archetypal sense of the Cosmic Tree and the Tree of Life. The tree symbol gives expression to the organic unity of the universe, but especially of the Earth in its integral reality. Obviously any damage done to the tree will be experienced throughout the entire organism. This could be one of our most effective ways of creating not simply conscious decisions against industrial devastation of the Earth but a deep instinctive repulsion to any such activity. This instinct should be as immediate as the instinct for survival itself.

A fourth symbol of great significance in this context is the Death-Rebirth symbol. This symbol is especially relevant to the cosmic process of continuing transformation. The industrial age came about by appeal to transformation symbolism, a passing from the old, the antiquated, the oppressive, the confining, into the new, the vital, the visionary, the liberating. Death-Rebirth symbolism as used by the modern industrial movement must now be turned from its destructive orientation toward a more integrating role.

These four symbols—the Journey, the Great Mother, the Cosmic Tree, and the Death-Rebirth symbol—experienced now in a time-developmental rather than a spatial mode of consciousness, constitute a psychic resource of enormous import for establishing ourselves as a viable species in a viable life system on the planet Earth.

Among the controlling professions in America, the educational and the religious professions should be especially sensitive in discerning what is happening to the planet and the value of these symbols in restoring a certain integrity to the human process. These professions present themselves as guiding our sense of reality and value at its ultimate level of significance. They provide our life interpretation. Education and religion, especially, should awaken in the young an awareness of the world in which they live, how it functions, how the human fits into the larger community of life, the role that the human fulfills in the great story of the universe, and the historical sequence of developments that have shaped our physical and cultural landscape. Along with this awareness of the past and present, education and religion should communicate some guidance concerning the future.

The pathos of these times, however, is precisely the impasse that we witness in our educational and religious programs. Both are living in a past fundamentalist tradition or venturing into New Age programs that are often trivial in their consequences, unable to support or to guide the transformation that is needed in its proper order of magnitude. We must recognize that the only effective program available as our primary guide toward a viable human mode of being is the program offered by the Earth itself.

Both education and religion need to ground themselves within the story of the universe as we now know it through our empirical ways of knowing. Within this functional cosmology we can overcome our alienation and begin the renewal of life on a sustainable basis. This story is a numinous revelatory story that could evoke not only the vision but also the energies needed for bringing ourselves and the entire planet into a new order of survival.

7

The UNIVERSITY

THE UNIVERSITY HAS A CENTRAL ROLE IN THE DIRECTION AND fulfillment of the Great Work. It seems appropriate, then, that we give some thought to the difficulties the university has experienced in recent times and the directions it might take in fulfilling its role in the twenty-first century.

The university can be considered as one of the four basic establishments that determine human life in its more significant functioning: the government, the religious traditions, the university, and the commercial-industrial corporations.

All four—the political, religious, intellectual, and economic establishments—are failing in their basic purposes for the same reason. They all presume a radical discontinuity between the nonhuman and the human modes of being, with all the rights and all inherent values given to the human. The other-than-human world is not recognized as having any inherent rights or values. All basic realities and values are identified with human values. The other-than-human modes of being attain their reality and value only through their use by the human. This attitude has brought about a devastating assault on the nonhuman world by the human.

Earlier human traditions experienced a profound intimacy with the natural world in all its living forms and even a deep spiritual exaltation in the religious-spiritual experience of natural phenomena. We have moved from this intimacy of earlier peoples with the natural world to the alienation of modern civilization. If some aesthetic appreciation remains, this seldom has the depth of meaning experienced earlier. Yet this presence to the natural world does occur with extraordinary power and understanding in persons such as Henry David Thoreau and John Muir and in many of the nature writers of the twentieth century, such as Aldo Leopold, Loren Eiseley, Edward Abbey, Edward Hoagland, Brenda Peterson, Barry Lopez, Terry Tempest Williams, Gary Snyder, David Rains Wallace, Annie Dillard, David Suzuki, Farley Mowat, and others too numerous to mention. Yet these writers have no role in forming the basic orientation of the contemporary university.

As now functioning, the university prepares students for their role in extending human dominion over the natural world, not for intimate presence to the natural world. Use of this power in a deleterious manner has devastated the planet. We suddenly discover that we are losing some of our most exalted human experiences that come to us through our participation in the natural world. So awesome is the devastation we are bringing about that we can only conclude that we are caught in a severe cultural disorientation, a disorientation that is sustained intellectually by the university, economically by the corporation, legally by the Constitution, spiritually by religious institutions.

The universities might well consider their own involvement in our present difficulties. Some of our most competent biologists in their comprehensive understanding of the biosystems of the planet, such as E. O. Wilson, Niles Eldredge, and Norman Myers, tell us that no devastation at this level has happened to the life systems of Earth since the termination of the Mesozoic Era some 65 million years ago (Wilson, *Biodiversity)*. The present, then, is beyond comparison with other historical changes or a cultural transition, such as that from the classical Mediterranean period to the medieval period or from there to the Enlightenment in Europe. Even the transition from the Paleolithic to the Neolithic Age in human cultural development cannot be compared to what is happening now. For we are changing not simply the human world, we are changing the chemistry of the planet, even the geological structure and functioning of the planet. We are disturbing the atmosphere, the hydrosphere, and the geosphere, all in a manner that is undoing the work of nature over some hundreds of millions, even billions of years. The genetic strains we have extinguished will never return.

Just what is involved in any full assessment of the disturbance of the planet need not be our concern here. Yet we might mention that in economics, the separation of the human economy from the Earth economy has been disastrous beyond measure. A rising gross human product with a declining gross Earth product is surely a contradiction. To preserve the integrity of the Earth economy should be the first purpose of any human economic program. Yet it would be difficult, until recently, to find a university where this first principle of economics is being taught. It is a strange thing to witness humans moving from suicide, homicide, and genocide to biocide and geocide under the illusion that they are improving the human situation.

Not only is this devastation of the natural world due to an industrial economy that is willing to wreck the entire planet for financial gain or some so-called improvement in the human condition. It is due also to the American Constitution, which guarantees to humans participatory governance, individual freedoms, and rights to own and dispose of property—all with no legal protection for the natural world. The jurisprudence supporting such a constitution is profoundly deficient. It provides no basis for the functioning of the planet as an integral community that would include all its human and other-than-human components. Only a jurisprudence based on concern for an integral Earth community is capable of sustaining a viable planet.

This legal status of rights for natural modes of being is especially needed now when the human has attained such extensive power over the functioning of the planet it possesses. So long as the American

Constitution in its present form and interpretation remains our ultimate referent in legal affairs, any equitable consideration of the natural modes of being of this continent will never be achieved. In the larger community of nations several steps have been taken to remedy this situation. The most impressive of these is the World Charter for Nature passed in the United Nations General Assembly in 1982. This charter states quite clearly that "Every form of life is unique, warranting respect regardless of its worth to man, and, to accord other organisms such recognition, man must be guided by a moral code of action." A similar attitude is expressed in the Earth Charter that is being prepared for presentation to the United Nations in the year 2002. This is a comprehensive document seeking to bring together social justice and sustainable development issues with environmental concerns.

The religious establishments are also seriously deficient in not teaching more effectively that the natural world is our primary revelatory experience. Emphasis on verbal revelation to neglect of the manifestation of the divine in the natural world is to mistake the entire revelatory process. Added to this is the excessive emphasis in Western religious traditions on redemption processes to the neglect of creation processes. This emphasis leaves us unable to benefit religiously from that primary and most profound mode of experiencing the divine in the immediacies of life.

A consciousness of this need for greater religious interest in the ecology issue led to a series of ten conferences held from 1996 to 1999 at Harvard University on the various religious traditions in their relation to the enviroment. This remarkable series of conferences brought together some eight hundred scholars and practitioners of the world's religions for reflection on the practical and theoretical resources of these traditions for mutually enhancing human-Earth relations. The papers from the conference are being published by Harvard University Press.

I mention economics, jurisprudence, and religion because these are among the subjects that are taught in our colleges and universities. An integral presentation of these subjects has not been given because of their commitment to the view that the nonhuman world is there fundamentally for the use of humans; whether economically, aesthetically, recreationally, or spiritually. For this reason the universities may be one of the principal supports of the pathology that is so ruinous to the planet.

Because of this basic attitude we consider that the more extensively we use the world about us, the more progress we are making toward some higher state of being. The vision of a transearthly status to be achieved by exploiting the natural world has driven us to ever more violent efforts toward this end. The ideal is to take the greatest possible amount of natural resources, process these resources, put them through the consumer economy as quickly as possible, then on to the waste heap. This we consider as progress—even though the immense accumulation of junk is overwhelming the landscape, saturating the skies, and filling the oceans.

It is important to note, however, four significant movements that have arisen to counter these directions. In the field of economics there is the Society for Ecological Economics established by Herman Daly and Robert Costanza. In jurisprudence is the emergence of the Earth Charter as a basis for recognition of the comprehensive Earth community. In the area of religion the Forum on Religion and Ecology arose from the three-year conference series at Harvard examining the various views of nature in the world's religious traditions. In education the greening of the university around the Tailloires Declaration is encouraging universities and their leaders to embody sustainable practices.

Yet there is still a deeper source of difficulty in the university. It lies in what are called the humanities, or liberal studies, as they are known. These supposedly, as humanist scholars tell us, provide for the expansion of the truly human quality of life. Yet this centering of value so extensively on the human distorts the place and role of the human in the structure and functioning of the universe. We fail to recognize that although the various components of the universe exist for each other, each exists primarily for the integrity of the universe. The human also, however noble in itself, exists for the integrity of the universe and for the Earth more than these exist for the human. Indeed the human depends upon the larger universe for its existence, its functioning, and its fulfillment. Within the order of the universe the planet Earth provides the efficient, final, material, and formal causes that bring the human into being, support the human in being, and lead the human to fulfillment.

The primacy of the universe over any part of the universe, and of Earth over any component of Earth, has been maintained earlier in our Western religious and cosmological traditions. The sacred community is primarily the universe community, not the human community. Whatever the deficiencies of medieval theological thinking it was clear that the entire universe is the primary value. The human belongs completely within the created order as a part of a more integral whole. As indicated by Thomas Aquinas, the most renowned of medieval theologians, "The order of the universe is the ultimate and noblest perfection in things" (Aquinas, SCG, bk. 2, chap. 46).

Even within the traditional theological context it could be said that what is done by the divine within the created order has for its supreme purpose the resplendence of the whole, not the resplendence of any single component of the whole. Only the whole has any integral meaning. Even the incarnation and redemption as these are presented within the Christian tradition must be considered as primarily for the good of the universe even though these have a certain immediate reference to the human. As was said at the time: "The whole universe together participates in the divine goodness and represents it better than any single being whatsoever" (Aquinas, ST, Q. 47, Art. 1).

Historically the break with this tradition took place at the time of the Great Plague that struck Europe in 1347–1349. This was a traumatic moment for the Western world. The deep aversion to the natural world that resulted has profoundly conditioned the Western cultural tradition ever since.

A definitive stage in this aversion came with René Descartes in the early seventeenth century. In a very real sense he *de*souled the Earth with his division of reality between mind and extension. In this perspective the nonhuman world was seen simply as mechanism. It was, however, a mechanism that could be, and even must be, exploited for human benefit.

For six centuries from the Great Plague and for more than three centuries from the time of Descartes, the aversion of the human from any intimacy with the natural world has increased in Western society, with the exception of the period of Romanticism from the late eighteenth century through the early nineteenth century. Scientists have insisted with ever-greater vehemence until recently that the universe can only be understood as the random action of minute particles with neither direction nor meaning. That we should have resisted such an interpretation given by scientists to their own discoveries is quite proper. That we should have permitted scientists to evoke in us a deep suspicion of the natural world is a matter of extreme regret.

We should have been able to provide our own interpretation of the scientific discoveries. It should have been obvious that our empirical inquiry into the structure and functioning of the universe was revealing a magnificent world beyond anything that we could have thought or dreamed. Any reasonable response is admiration, awe, and even a certain foreboding at the deeper mysteries presented in such an overwhelming reality. We might even consider that the emergent universe, in the sequence of its unfolding, is providing us with a new revelatory experience of whatever is the origin from which it emerges.

For this experience we do not need telescope, microscope, or scientific analysis. Yet with these instruments of intimacy with the universe we do have a new understanding of the sequence of transformations through which the evolutionary process has passed in becoming what we observe in the present. If the religious experience were simply some naive impression of the uninformed it would not have resulted in such intellectual insight, such spiritual exaltation, such spectacular religious ritual, or in the immense volume of song and poetry and literature and dance that humans have produced.

Few indeed, it seems, are those whose vision of the stars, the ocean, the song and flight of birds, the exquisite form and activities of the various animal species, or the awesome views of the mountains and rivers and valleys does not evoke some sense of an inner spontaneity, a guiding principle, a consciousness, a transmaterial presence manifested throughout the material embodiment, an ordering principle observed in any living being that enables the complexity of the DNA in the genetic process to function in some coherent fashion. While no sense faculty can experience it directly and no equation can be written to express it, our immediate perception tells us that there is a unifying principle in the acorn that enables the complex components of the genetic coding of the oak tree to function as a unity—send down roots, raise the trunk, extend the branches and put forth leaves and fashion its seeds, then to nourish all this by drawing up tons of water and minerals from the Earth and distributing them throughout the entire life system. That such a vast complexity of functioning should have some unifying principle, known traditionally as the "soul" of the organism, is immediately evident to human intelligence.

Since this is not the occasion to argue the case for the psychic or the spirit or the soul dimension of living organisms, I will only indicate that my generation has been an autistic generation in its inability to establish any intimate rapport with the natural world. This mental deficiency has brought us into the terminal phase of the Cenozoic Era in the geobiological story of Earth development. Our present need is to know just how to move out of this alienation of the human into a more viable mode of presence to the natural world.

Here I propose that the religions are too pious, the corporations too plundering, the government too subservient to provide any adequate remedy. The universities, however, should have the insight and the freedom to provide the guidance needed by the human community. The universities should also have the critical capacity, the influence over the other professions and the other activities of society. In a special manner the universities have the contact with the younger generation needed to reorient the human community toward a greater awareness that the human exists, survives, and becomes whole only within the single great community of the planet Earth.

If the central pathology that has led to the termination of the Cenozoic is the radical discontinuity established between the human and the nonhuman, then the renewal of life on the planet must be based on the continuity between the human and the other than human as a single integral community. Once this continuity is recognized and accepted, then we will have fulfilled the basic condition that will enable the human to become present to the Earth in a mutually enhancing manner.

In this new context every component of the Earth community would have its rights in accord with the proper mode of its being and its functional role. In each case the basic rights would be for habitat and the opportunity of each being to fulfill its role in the natural systems to which it belongs. Humans would be obliged to respect these rights. If such concerns were not under discussion in the eighteenth century when the American Constitution was being written, they must be the central issue in any present discussion of the legal context of our society. The critical mission of the university law schools is to address these issues in a depth that has not yet manifested. A more expanded basis for jurisprudence seems to be indicated. A beginning has been made by Justice William O. Douglas in *A Wilderness Bill of Rights,* published as long ago as 1965. There we find a remarkable affirmation of the need to establish legal status for the natural world.

Even beyond the Earth, the sense of community would extend throughout the entire universe seen as a single coherent community that has emerged into being with a total dependence of each component on all the others. Indeed, we need to think of the universe as the supreme norm of reality and value, with all component members of the universe participating in this context, each in accord with its own proper role.

In this setting the universe would become the primary university, just as the universe is the primary lawgiver, the primary economic corporation, the primary scientist, the primary technologist, the primary healer, the primary revelation of the divine, the primary artist, the primary teacher, and indeed the primary source, model, and ultimate destiny in all earthly affairs. Throughout our human intellectual development we are totally dependent on what the universe communicates to us in an earlier stage through immediate observation and in this later stage through all those instruments of observation that we have devised. Through these instruments of observation we enter profoundly into the most hidden realms of phenomenal existence itself while at the same time these hidden realms enter into our own minds. It is a reciprocal relationship. We are touched by what we touch. We are shaped by what we shape. We are enhanced by what we enhance.

The human university would be the context in which the universe reflects on itself in human intelligence and communicates itself to the human community. The university would have the universe as its originating, validating, and unifying referent. Since the universe is an emergent reality the universe would be understood primarily through its story. Education at all levels would be understood as knowing the universe story and the human role in the story. The basic course in any college or university would be the story of the universe.

This story can fulfill its role only if the universe is understood as having a psychic-spiritual as well as a physical-material aspect from the beginning. This should not be difficult since we know what something is by its appearance and by what it does. We know a mockingbird by the variety of its songs, by its size, by the slate gray color of its feathers and by the white patch on its wings and the white feathers in its tail. Since the universe brings us into being with all our knowledge and our artistic and cultural achievements, then the universe must be an intellect-producing, aesthetic-producing, and intimacy-producing process.

These qualities that we identify with the human are also qualities that we observe throughout the natural world. Even at the level of the elements we observe self-organizing capacities, also the capacity for intimate relationships. These reveal astounding psychic abilities. These are so impressive that we must consider that modes of consciousness exist throughout the universe in a vast number of qualitatively diverse manifestations. Above all we discover that every being has its own spontaneities that arise from the depths of its own being. These spontaneities express the inner value of each being in such a manner that we must say of the universe that it is a communion of subjects, not a collection of objects.

Precisely in this intimate relationship with the entire universe we overcome the mental fixation of our times expressed in the radical division we make between the human and the other-than-human. This fixation that I have described as an unfeeling relation of the human to the natural world is healed in its deepest roots as soon as we perceive that the entire universe is composed of subjects to be communed with, not primarily of objects to be exploited. This communion experience is, I believe, universal. It can be observed in the immediate reaction of almost anyone who simply looks at the ocean at dawn or sunset or at the heavens at night with all the stars ablaze, or who enters a wilderness area with its foreboding as well as its entrancing aspects.

In every phase of our imaginative, aesthetic, and emotional lives we are profoundly dependent on this larger context of the surrounding world. There is no inner life without outer experience. The tragedy in the elimination of the primordial forests is not the economic but the soul-loss that is involved. For we are depriving our imagination, our emotions, and even our intellect of that overwhelming experience communicated by the wilderness. For children to live only in contact with concrete and steel and wires and wheels and machines and computers and plastics, to seldom experience any primordial reality or even to see the stars at night, is a soul deprivation that diminishes the deepest of their human experiences.

Here I propose that the universities need to teach the story of the universe as this is now available to us. For the universe story is our own story. We cannot know ourselves in any adequate manner except through an account of the transformations of the universe and of the planet Earth through which we came into being. This new story of the universe is our personal story as well as our community story.

We have moved from a sense of time in which the universe revolves simply in ever-renewing seasonal cycles into a universe that has emerged into being through a sequence of irreversible transformations, even while it is also revolving in an ever-renewing sequence of seasonal changes. Our greatest single need is to accept this story of the universe as we now know this as our sacred story. It could be considered as the most magnificent of all creation stories. This story does not diminish, it rather enhances the earlier story that we have through the Book of Genesis. That story was related to the ancient Mesopotamian stories of the universe. Our new story is attained in a more empirical manner and with new instruments of observation.

We now know ourselves as genetically related to every other living being in the universe. Only through this story are we able in any integral manner to overcome our alienation from the natural world about us. We are finally able to understand just why our own well-being is dependent on the well-being of Earth. Yet even when we know this with such depth of understanding we still find it difficult to rethink economics, law, religion, and education within this scientific context. Our universities seem caught in a fixation from which they cannot escape even when these prior cultural forms are proving unable to prevent the devastation of the planet.

Such fixation on our existing cultural forms remains, apparently, the only context for survival that the universities can appreciate. The difficulty is not exactly in the cultural forms but in the inability to expand understanding of how these cultural forms function in this new context. The difficulty is also in a misunderstanding or excessive emphasis on some phases of these cultural forms—such as, for instance, the religious emphasis on redemption to the neglect of creation. So too our inability to understand that these prior cultural forms will enter a more expansive phase of their existence within this new context than they ever had in previous times.

The urgency of moving into the new situation would not be so great if the devastation of the planet were not so overwhelming. As long as we live and have our values and do our educating within this prior context we will not be able to establish a deeper understanding with those who have.

While our universities have gone through many transitions since they first came into being in the early medieval period, they have never experienced anything like the transition that is being asked of them just now. The difficulty cannot be resolved simply by establishing a course or a program in ecology, for ecology is not a course or a program. Rather it is the foundation of all courses, all programs, and all professions because ecology is a functional cosmology. Ecology is not a part of medicine; medicine is an extension of ecology. Ecology is not a part of law; law is an extension of ecology. So too, in their own way, the same can be said of economics and even the humanities.

There have been stages when the Western university was dominated by theology as the queen of the sciences. There have been periods when the universities were dominated by humanistic concerns. There have been times when the university was dominated by mechanistic science, engineering, or business. The new situation requires that the university find its primary concern in a functional cosmology. Such a functional cosmology can exist, however, only within a university where the spirit dimension of the universe as well as its physical dimension is recognized.

The transformation of human life indicated in this transition from the Cenozoic to the Ecozoic Era affects our sense of reality and values at such a profound level that it can be compared only to the great classical religious movements of the past. It affects our perception of the origin and meaning of existence itself. It might possibly be considered as a metareligious movement since it involves not simply a single segment of the human community but the entire human community. Even beyond the human order, the entire geobiological order of the planet is involved.

At this opening period of the third millennium of our times there are choices to be made in every phase of human life. The immediate decision is whether any of our basic institutions—government, religious establishments, universities, or corporations—can mitigate their attachment to the terminal phase of the Cenozoic, or whether any one of these or all these might make this change in its full order of magnitude.

The universities must decide whether they will continue training persons for temporary survival in the declining Cenozoic Era or whether they will begin educating students for the emerging Ecozoic. Already the planet is so damaged and the future so challenged by its rising human population that the terms of survival will be severe beyond anything we have known in the past. We have not thought clearly or behaved properly in the twentieth century. We are now caught in a mind-tormenting ambivalence. We have such vast understanding of the universe and how it functions, and yet we manifest such inability to use this knowledge beneficially either for ourselves or for any other mode of earthly being. While this is not the time for continued denial by the universities or for attributing blame to the universities, it is the time for universities to rethink themselves and what they are doing.

8

ECOLOGICAL GEOGRAPHY

GEOGRAPHY IS AN INTEGRATING STUDY OF THE EARTH IN ITS comprehensive extent as well as in its various regional integrations. It does, however, need the other sciences for the specific data that they supply. Geography needs geology for understanding the structure of the land and the formation of the mountains and the rivers; biology for the study of life-forms; meteorology for understanding the weather of regions that make a region suitable for particular forms of vegetation; hydrology for understanding the rivers, lakes, and aquifers, and forestry for the study of the wooded areas.

While geography provides a comprehensive context for understanding the functioning of the Earth in its larger structure, it is even more useful in appreciating the integral functioning of the various regions into which the planet is divided. In this manner it provides the context for ecological understanding.

Earth, we might say, is a single reality composed of a diversity beyond all understanding or description. This diversity in its arctic and tropical regions, its oceans and its continents, in its mountains and valleys, its forests and deserts, its rivers and their floodplains, all give to Earth both its endless wonder and its functional integrity. These landscape features and these living forms have come into being as some self-woven tapestry or some self-composed symphony or some self-designed painting. To experience this wonder and to enter into intimate relations with the various life communities of these regions seems to be the high purpose of human presence on the Earth.

Prior to the emergence of humans the Earth, with the rise and fall of continents, their coming together and their drifting apart, with the sequence of geological layers in the formation of the continents, with their volcanic eruptions, forests, and wildlife, with all these apparently destructive experiences, has remained coherent and creative throughout the vast period of its development.

That the geological structure of Earth and the number of life species should be so vast, so diverse, and their interaction so intimate is the wonder of the known universe. No other planet has, so far as we can determine, anything beyond minimal indications of any possible life, much less such magnificent forms or such diversity of species. Even after all these years of scientific inquiry into the structure of the planet and its life systems, our knowledge of the forces creating the physical contours of the planets, the sequences of climate change, and the disposition of species inhabiting this planet is still quite limited. While we know more than a million and a half species we estimate that there are at least ten to twenty million or even more species occupying the Earth. The interaction of these species with their geographical location constitutes a major phase of land formation.

Then came the human, a being with a genetic mandate to shape a cultural mode of being of its own design but with a dependence on all the other forces shaping the continents. This power of self-shaping is possessed by all living beings, but in relation to other modes of being and within controlled limits, which have come to be designated as the species "niche." Geographical context is among the most powerful determinants in enabling the human to establish its place in the community.

The variety of regions where a single species might dwell was vastly expanded by the type of intelligence granted to humans. While intelligence enabled humans to adapt more extensively to the outer world, this same intelligence gave to humans the capacity to fashion for themselves a variety of inner cultural worlds. These cultural realms would constitute not only the inner determination of the human, they would also determine the relation of humans to the other modes of being. In a special manner humans, in their adaptation to geographical environment, would require a conscious understanding of the place of their dwelling.

Several million years were needed before humans could fully shape their identity within the geographical context in which they found themselves. Then it required another long period of time for humans to relate effectively to the other members of the surrounding community. This capacity took its most significant step with the cultivation of the land and the confining of animals. From that beginning some ten thousand years ago this capacity increased, at first gradually, then with ever-accelerating impact on the biosystems of the community. Eventually the achievement of a truly human way of life came to be judged by the extent of control over the geological structure, the vegetation, and the wildlife-forms in the region and their use for human purposes.

What also distinguished the human mode of being was the sense of spirit powers present throughout the geographical region. The rivers and moutains were not simply physical forms, they were spirit powers to be reckoned with. The sense of relating to spirit powers identical with the topography of the region established one of the specific differences in the human adaptation to regional context and to other life-forms. It also provided the intense emotional attachment for human communities to the place of their dwelling.

In recent times what industrial civilizations have failed to realize is that in the particular place of their dwelling the well-being of the Earth was a necessity for their own well-being and fulfillment. The attitude that the Earth existed for utilitarian human purposes became progressively severe as commitments to the individual rights of humans were enacted into political constitutions with no corresponding rights being recognized for the other components of the natural life community. Legal enactments gave to humans what were designated as "property rights" over land and whatever existed on the land. Such enactments provided the basis for occupation and unrestrained exploitation of the land. Western civilization, dominated by a cultural arrogance, could not accept the fact that the human, as every species, is bound by limits in relation to the other members of the Earth community.

While refusal of any other members of the Earth community to accept limits might quickly lead to extinction of the disturbing species, humans found that they could, for a period of time, subvert the forces that might normally lead to their own extinction. What humans could not do was to avoid the degradation in their own mode of being that occurs as soon as they prevent the other members of the community from fulfilling their role in the larger Earth community. Only gradually have modern humans in the Western cultural tradition begun to realize that we have a profound need for the well-being of other species if we are to experience any well-being or fulfillment in ourselves.

In these opening years of the twenty-first century we need to renew our intimacy with our local bioregion and with the North American continent but also with the planet Earth itself, in its comprehensive extent and in the diversity of its component regions. To accomplish this intimacy in some integral manner we require a study that would fulfill the ideal of a "total earth science" that was spoken of so frequently by Robert Muller, a former adviser to several secretaries-general of the United Nations from the 1950s until the 1970s. This phrase, total earth science, seems to have the comprehensive extent and the precision in statement needed in designating an area of understanding that has never been given its proper identity or its proper place in our educational program.

This lack of attention to an integral Earth study is one reason for the difficulty humans experience in finding their place within the dynamics of the Earth. Some beginnings in understanding have been made in studies of the Earth in terms of the Gaia hypothesis. This refers to the Earth as having the capacity for *homeostasis;* that is, for comprehensive inner adjustment and self-regulation in response to changes in the outer world. This concept has led to macrophase biological studies of the landsphere, the watersphere, the airsphere, the life sphere, and the mindsphere as the five macrophase components of the Earth.

This way of thinking about the planet provides an integrating context for all our particular studies. It provides a way of understanding and managing the complexity and tensions that exist amid the vast array of forces that enable the Earth to be the wonder planet that it is. It enables the human community to begin thinking more adequately of its own role. In recent times our human role has been profoundly altered due to the powers attained by the human community through the sciences and technologies that have been developed in the twentieth century. Humans are now altering the planet on a scale that can be compared with the glacial periods in the influence that they are now having on the planet. Our capacity to extinguish lifeforms has even been compared with the forces that terminated the Mesozoic and introduced the Cenozoic eras in Earth history some 65 million years ago.

It is difficult to appreciate the full extent of the power possessed by industrial civilizations to disrupt the integral functioning of the life systems of the Earth. A new geobiological age has been introduced. This requires new perspectives in those studies that are concerned with the integral functioning of the Earth: geology, chemistry, biology, and those other studies of the Earth and its manner of functioning. In all these areas the natural systems have been profoundly affected by the new powers of the human community.

Here we must once again refer to the fact that humans are different from other species in establishing their self-identity and in identifying their role in relation to the other components of the Earth community. To some extent this entire book can be considered an effort to identify the role of the human community in relation to the other components of the planet. While this effort can be understood as the search for the proper niche of the human, it is a special form of niche. The other-than-human species, through their genetic endowment, discover their survival context with only limited disturbance of the larger complex of life systems. They find their niche quickly, or else they perish. A certain stability eventuates. A new equilibrium comes into being, a functioning relation of things with one another.

The difficulty with humans is that we are genetically coded toward a further transgenetic cultural coding whereby we establish ourselves in our specific mode of being, a cultural mode of being that we hand on to succeeding generations by education. Through this cultural mode of being we also establish our niche in the Earth community. Our place in the community is more extensive in its habitat region than is the case with other modes of being. Whereas other life-forms generally survive only within a limited bioregion, we can establish our human presence almost anywhere on the planet.

In some sense the human refuses to accept any particular niche, for the basic function of a niche is to set limits to the activity of a species. In this sense the human refuses to accept limits imposed from without or even from within its own being. By bringing humans into existence the Earth has created a supreme danger to all other components of the Earth community because the human can invade the region of other species with a unique range of freedom.

Survival of any group of living beings in relation to other groups depends on the recognition of limits in the actions of each group. This law of limits is among the most basic of all cosmological, geological, or biological laws. It is particularly clear in the case of biological forms. In the Hindu world this law of limits is recognized as *rita* in the cosmological order or as *dharma* in the moral order. In the Chinese world it is *tao* in an earlier phase or *li* in the later neo-Confucian period. In the Greek world it is *dike* as the order of justice or *logos* establishing the intelligible order of the universe. Yet in the modern world this sense of limits imposed by the natural functioning of the universe has to some extent been overridden, at least in a temporary manner, by industrial processes created by humans.

The general law is that every species should have opposed species or conditions that limit them so that no single species or group of species would overwhelm the others—something that would assuredly happen if even a bacterium were permitted to reproduce without limitation over a period of time. The law of limits is what makes the functional rapport between the various life-forms an urgent necessity.

That is the difficulty for humans. We must self-limit. We have such an extensive range of abilities in relation to the other components of the planet that we seem not to know where to place the limits on our actions. Or perhaps we are simply unwilling to limit ourselves by deliberate decision. To some extent this derives from our partial emergence out of the controls of instinct when we acquired the capacity for intelligent thought. In the twentieth century we have been so entranced with our evolutionary origins and the long series of transformations that have brought us into being that we are more attracted to cosmogenesis than to cosmos. Even as regards Earth we are more committed to history than to geography, more committed to time than to space. History is endless. Place is limited.

We are so impatient with our given place in the universe that some persons are totally committed to discovering how we can get beyond Earth. We have indeed been out in space, but some are under the illusion that we have been off Earth. In reality humans have never been off Earth. We have always been on a piece of Earth in space. We survive only as long as we can breathe the air of Earth, drink its waters, and be nourished by its foods. There is no indication that as humans we will ever live anywhere else in the universe. Place, too, is continuously being transformed but only within its own possibilities.

Our entire industrial system can be considered as an effort to escape from the constraints of the natural world. We have created an artificial context for our existence through mechanical invention and the extravagant use of energy. In this process we have so violated the norms of limitation, so upset the chemical balance of the atmosphere, the soil, and the oceans, so exploited the Earth in our use of fossil fuels, that we are devastating the fertility of the planet and extinguishing many species of wildlife. We no longer live within the organic, ever-renewing world that is the natural context of our existence.

When we awaken to a realization that the industrial world, as now functioning, can exist for only a brief historical period, we might begin to consider just how we can establish a more sustainable setting for our physical survival and personal fulfillment. We must, obviously, turn from our exploitation of the natural world to consider once again just how the planet functions and where we belong in relation to the other components of the planet. Since we do not function primarily by instinct, we can do very little until we have some idea of how the life systems of the Earth function in producing the food and shelter and the energies we need. In some sense this is a recovery process, since in our agricultural phase we had an abundance of knowledge of how the Earth functions in its various bioregions.

Now, however, we need a much more comprehensive type of understanding and a more extensive human adaptation to the various bioregional contexts of our dwelling. While we need this intimate acquaintance with the organic functioning of our local region, we also need a larger sense of the Earth. We have become so conscious of the planetary context of our lives that we cannot completely withdraw into the local region.

However resistant to the restraints inherent in their nature, humans in the natural order of things belong to, are possessed by, and are subject to the geographical place where they reside. Yet through technological skills humans have become less dependent on their immediate geographical region. We have come to consider that we become more human the more extensively we withdraw from any dependency on our bioregion. Our present alienation is such that we have little concern for where our food comes from—whether it is grown in North America, Africa, Australia, or South America. Our clothing could be made from raw materials grown in one country, shipped to another to be tailored, and then shipped abroad to be sold somewhere, anywhere. There is little or no relation to the fields that grow our food, to the streams that provide our water, to the woodlands that surround us, or to the regional flowers or fauna. Moreover, there is frequently exploitation of the labor of those who grow our food and make our clothes.

This psychic world of no attachment, no intimacy, is also the world of no fulfillment. There is effectively no feeling of intimacy with our place. While we expect our place to give itself to us, we have no sense of giving ourselves to our place. Regional feeling or understanding has become an irrelevant intellectual discipline. It is given little attention in the education of children. If geography is included, it is political or economic geography for human purposes, not for the purposes of the larger Earth community. Yet as population increases and available space on the planet becomes more limited, the study of Earth and its regions becomes more critical. Economic geography needs to discover where the living resources of the Earth are located in each bioregion, how abundant are these resources, how they are best sustained in their capacity for unlimited renewal.

The founder and most distinguished of the modern geographers, the one who designed new ways of thinking about and recording the structure and functioning of the planet, was Alexander von Humboldt (1769–1859). Based on his studies in South America he provided an overview of the geological structure, the climatic conditions, and the vegetation of that region. Then he ventured into wider studies of the geological structure and functioning of the European, Central Asian, and American worlds. These studies he published in twenty-two volumes. His best-known work, however, is *Cosmos,* published in English in five volumes (1845–1862), the first great treatise on modern geography.

Humboldt was followed by the French geographer Jean-Jacques-Elisée Reclus (1830–1905) with his central work *La Terre* (1868–1869). Eduard Suess (1831–1914), a contemporary of Reclus, wrote as his principal work, *Das Anlitz der Erde (The Face of the Earth)* in five volumes published in English in 1883–1909. These early geographers were followed by such scholars as Friederich Ratzel (1844–1904) and by a series of French scholars who were mostly interested in regional geographies.

These are the sources from which geographical studies emerged in America at the end of the nineteenth century. Isaiah Bowman (1878–1950), director of the American Geographical Society for twenty years, had a primary interest in geography in its relation to the social sciences. Among his special interests were the Andes regions of southern Peru and northern Chile. Carl Sauer (1880–1975) did special studies of the desert regions of the southwestern areas of the North American continent. His studies included the human geography of the American Indian and also the early agriculture of Mexico. These terms, *ecology* and *geography,* were brought together as early as 1923 in H. H. Burrows's presidential address to the Association of American Geographers, which was entitled "Geography as Human Ecology." Ellsworth Huntington (1876–1947) did a comprehensive study of the Earth and its influence on human formation in his monumental study *Climate and Civilization,* which was published in 1915.

Since this last part of the nineteenth century and the first half of the twentieth was the high period in the imperial dominance over the Earth by the Western powers, it was also the time when support for political-military dominance of the world was sought in geographical considerations. This led to the views of the American naval officer

Alfred Thayer Mahan (1840–1914) concerning sea power, set forth in his study of *The Influence of Sea Power upon History, 1660–1783*. Global strategies based on the land came into being through Sir Halford John Mackinder (1861–1947). He established geographical studies in England as an academic discipline at the London School of Economics while developing his theory of the land basis of power in Eurasia as the geographical heartland of the Earth and of human history, a thesis that he presented in an essay in 1904.

Although these uses of geographical data are immensely important in understanding past history and the forces that have brought us to our present world situation, we must move on to other concerns. We are now more concerned with the influence of human civilizations on the climate than with the climate's influence on human civilizations.

While cultural geography, economic geography, political geography, and military geography have served the purposes of human exploitation of the planet, the time has come to study the Earth for the purposes of the Earth. The well-being of the Earth depends to an extensive degree on our understanding of the planet in its global extension, in its bioregional diversity, and in the intimacy of the component parts in the whole. We depend on this understanding of the Earth in all its diversity if we are to know how humans are to be present to the planet in some mutually enhancing manner. Such understanding is the proper role of ecological geography. If this study were properly developed then a great advance will have been made toward achieving a viable planetary community.

Our present concern for the human venture in its relation to the natural systems of the planet is sometimes referred to as the "human *problematique*" or as the "global *problematique*." In either case the issue is the same. For we have a comprehensive Earth issue as well as a vast diversity of human issues with which to deal. In both instances some sense of the planet Earth as the intimate place of our dwelling needs to be fostered. The phrase "world *problematique*" has been extensively used in discussions such as those begun in the 1960s by the Club of Rome in the first extensive survey of our human future in light of the extreme demands we are making on the resources of the planet. This was the issue dealt with in the *Limits to Growth* (Meadows et al., eds., 1972), a study of the resources of North America and of the larger planet for purposes of long-term planning for a future human economy. Even the title of this book aroused antagonism throughout the commercial world as well as in the academic world. Its thesis—that we are imposing burdens on the Earth that are beyond Earth's carrying capacity—remains a powerful critique of the industrial-commercial-financial world of the twentieth century. Another such study, *Global 2000: A Report to the President,* was requested by President Carter in 1979, the last year of his presidency, for guidance in establishing long-term economic strategies for this country. The report came to the same conclusion and evoked similar resentment. While President Carter was aware that human intrusion into the planet's functioning was reaching the Earth's limits of sustainability, the succeeding president, Ronald Reagan, objected to the report and suppressed its printing by the Government Printing Office.

In resolving our present impasse, ecological studies are providing much-needed guidance. Indeed the term ecology is becoming a prefix that can be attached to almost any science or to any human activity. So we find ecological studies in their relation to law, economics, education, literature, ethics, and a large diversity of other aspects of the human project. In the future none of these activities will be able to proceed without a better geographical context of understanding the planet.

The more humanistic realms of poetry and the natural history essay are important to establish the emotional-aesthetic feeling for the wonders of the natural world and to awaken the psychic energies needed for dismantling our present destructive technological-industrial-commercial structures and creating a more benign mode of economic survival for the entire Earth community. But these humanistic insights are themselves ineffective unless they are enhanced by a more thorough understanding of the identifying features and intimate modes of functioning of bioregional communities.

The understanding of the Earth that we are indicating needs to be something more than a composite of these multiple ways of viewing the planet. For the present this idea of a total or integral Earth study seems implicit in what we presently designate as ecology. Another term coming into use is *Earth literacy*, as a basic context for educational programs from the earliest years through professional levels. Earth literacy is being fostered especially by educators such as David Orr of Oberlin College and Chet Bowers of Portland State.

Each of these terms has its own special value. My own expectation is that the study of ecological geography will have a significant role to play in the future as one of the most effective disciplines leading to an integral human presence to the larger Earth community. This presence will occur, however, only when the study of the Earth gives rise to an appreciation such as that given expression by John Muir in his writings on the Yosemite Valley in California.

What is needed is geography as an intimate study. Just as there is an affection between animals and humans, so there is an affection that passes between the region and human appreciation. Nothing escapes the role of intimacy. There is such a thing as considering the curvature of space as an intimacy of the universe with every being in the universe. So with the bioregion, there is an intimacy that brings to fulfillment both the region and its human presence. The region responds to the attention it receives from the various members of the community.

This feeling relationship with the Earth intensifies as we grow in familiarity with the region. As described by Barry Lopez in his essay on American geographics, "The more superficial a society's knowledge of the real dimensions of the land it occupies becomes, the more vulnerable the land is to exploitation, to manipulation for short-term gain. The land, virtually powerless before political and commercial entities, finds itself finally with no defenders. It finds itself bereft of intimates with indispensable, concrete knowledge" (Lopez, *About This Life,* p. 137).

Indigenous peoples know their region. They must know where the food is, where water is available, where firewood is found, where the medicinal plants are, where the trees grow that furnish the poles for their tents or the wood for their fires. Our studies in what we call ecology must lead to such intimacy with our natural surroundings. Only intimacy can save us from our present commitment to a plundering industrial economy.

9

ETHICS *and* ECOLOGY

IN APRIL OF THE YEAR 1912 THE *TITANIC*, ON HER MAIDEN voyage across the Atlantic, struck an iceberg and went down at sea. Long before the collision those in command had abundant evidence that icebergs lay ahead. The course had been set, however, and no one wished to alter its direction. Confidence in the survival capacities of the ship was unbounded. Already there were a multitude of concerns in carrying out the normal routine of a voyage. What happened to that "unsinkable" ship is a kind of parable for us, since only in the most dire situations do we have the psychic energy needed to examine our way of acting on the scale that is now required. The daily concerns over the care of the ship and its passengers needed to be set aside for a more urgent concern, the well-being of the ship itself. Here is where macrophase concerns in one context become microphase concerns in another context. Passenger concerns in the situation of the *Titanic* needed to give way to a macrophase decision about the ship itself.

Now our concerns for the human community can only be fulfilled by a concern for the integrity of the natural world. The planet cannot support its human presence unless there is a reciprocal human support for the life systems of the planet. This more comprehensive perspective we might identify as macrophase ethics. This is something far beyond our ordinary ethical judgments involving individual actions, the actions of communities, or even of nations. We are presently concerned with ethical judgments on an entirely different order of magnitude. Indeed, the human community has never previously been forced to ethical judgments on this scale because we never before had the capacity for deleterious action with such consequences.

As indicated by Brian Swimme in *The Hidden Heart of the Cosmos,* humans, through our scientific insight and our technological skills, have become a macrophase power, something on the level of the glaciations or the forces that caused the great extinctions of the past. Yet we have only a microphase sense of responsibility or ethical judgment. We need to develop a completely different range of responsibility.

It is not easy for us to move beyond those basic points of reference that have guided our way of life in former times, for these have given us our human identity and directed our religious and cultural traditions over the past millennia. These traditions have determined our language, our intellectual insights, our spiritual ideals, our range of imagination, our emotional sensitivities. Yet these classical traditions of the Eurasian and American worlds are all proving inadequate in dealing with the disintegrating influence we are now having on the life systems of the Earth. Yet we experience a kind of paralysis in our critical judgment of what is happening and what we need to do at this time to avoid an extensive crash of the biosystems of the planet. Much of the wisdom of the past becomes inadequate in the present.

One of the difficulties is in our language. Our traditional European languages express the anthropocentrism of our past orientation. Our Western imagination is filled with images derived from these same sources. Our traditional spiritual values are disorienting by their insistence on the unsatisfactory nature of the existing order of things and the need for relief by reference to some transearthly experience.

Religious persons are constantly asserting the high spiritual nature of the human against the lack of any spiritual dimension of the natural world. All earthly affairs are considered microphase concerns relative to the spiritual concerns that determine our destiny in some other transcendent world.

In recent times as our religious traditions have diminished in their influence over our lives, it is the human that dominates the scene. Nothing is considered superior to individual or community values. Our legal system fosters a sense of human rights, with other-than-human beings having no inherent rights. Our economics is based on our mechanistic exploitation of the Earth in all of its geobiological systems. Commercial rights to profit prevail over urgent needs of natural systems for survival. Disengagement from such exclusive commitments to human exploitation requires an ethical stance and a courage of execution seldom found in contemporary human societies.

In evaluating our present situation, E. O. Wilson of Harvard University tells us: "In the end it will all come down to a decision of ethics, how we value the natural world in which we have evolved and now—increasingly—how we regard our status as individuals" (Wilson, *Biodiversity*, p. 16). Paul Ehrlich, professor of biological sciences at Stanford University, has suggested that "scientific analysis points, curiously, toward the need for a quasi-religious transformation of contemporary cultures" (in Wilson, *Biodiversity*, p. 26).

The traditional religious orientation of Western society has made us vulnerable to superficial attitudes toward the difficulties we experience. When in a position of great danger we are prone to believe we will be saved by some transearthly intervention within the functioning of the planet. Such intervention will provide a remedy in the present as it has, supposedly, done on so many occasions in the past. The most glowing presentation of such a future is to be found in the apocalyptic literature of the Bible with its vision of the thousand-year period at some future time when the blessed will attain the glory of the first resurrection (Revelations 20:2). Such a transformation implies a glorious existence when sorrow will be eliminated, justice will reign, and peace will pervade the land.

Even when the religious dimension of the millennial search gave way to a humanistic life attitude, the sense of living in a radically unsatisfactory world remained a central fact in our Western consciousness. We have rarely felt at ease amid the spontaneities of the natural world. We feel we deserve a better world. We must find our fulfillment in some transformed earthly condition. We find increasing difficulties in accepting life within the conditions that life has granted us.

The means of dealing with this situation over the greater part of our Western history was through some inner discipline that would enable us to absorb the stress inherent in our earthly existence. Then, under the guidance of Francis Bacon in the seventeenth century, we began to envisage the possibility of understanding and controlling the processes of nature and thereby bringing about relief from the human condition through our own efforts. Nature began to be looked at both as an obstacle to be overcome and as a resource to be exploited. The ideal of a transformed society continued to be energized by a vision of the millennium. Only now the millennial experience was to be sought not through divine intervention but through scientific insight, technological skills, and commercial negotiation.

We know the story of the formation of the modern world, the dominant intellectual framework and its beginnings in the seventeenth century when Descartes established an absolute separation of the spiritual and the material worlds. Later in the seventeenth century Newton provided a view of the physical universe that came to dominate the Western mind until the time of Albert Einstein and Max Planck at the beginning of the twentieth century. This mechanistic view of the world encouraged the growth of technological invention and industrial plundering, culminating in the 1880s when the electronic and chemical research centers were established, scientific technologies were advanced, and the first modern commercial and industrial corporations were formed. The objective was to make human societies as independent as possible from the natural world and to make the natural world as subservient as possible to human decisions. Nothing was to be left in its natural state.

Only now can we appreciate the consequences of this effort to achieve human well-being in a consumer society by subduing the spontaneities of the natural world with human manipulation. We begin to realize that the devastation taking place cannot be critiqued effectively from within the traditional religions or humanist ethics. Nor can it be dealt with from within the perspectives of the industrial society that brought it about.

We find ourselves ethically destitute just when, for the first time, we are faced with ultimacy, the irreversible closing down of the Earth's functioning in its major life systems. Our ethical traditions know how to deal with suicide, homicide, and even genocide; but these traditions collapse entirely when confronted with biocide, the extinction of the vulnerable life systems of the Earth, and geocide, the devastation of the Earth itself.

We have a radically new *problematique.* To appreciate this fully we must understand that the misuse of our scientific-technological powers has not itself come ultimately from the scientific tradition, although this is the general accusation made against the empirical inquiry into the functioning of the natural world. The danger and the misuse have come ultimately from the deficiencies of the spiritual and humanist traditions of Western cultural development. These traditions themselves have alienating emphases. Both our religious and our humanist traditions are primarily committed to an anthropocentric exaltation of the human.

Consistently we have difficulty in accepting the human as an integral part of the Earth community. We see ourselves as a transcendent mode of being. We don't really belong here. But if we are here by some strange destiny then we are the source of all rights and all values. All other earthly beings are instruments to be used or resources to be exploited for human benefit. Now, after centuries of plundering the

Earth for our own advantage, we begin to reflect on who we are and what has happened both to the planet and to ourselves. A sudden reversal is taking place even while our bright, new, antiseptic, mechanical world is finding its fulfillment in the global range of its activities. The inescapable question arises: what is gained and what is lost? It is now a question of gain or loss on an absolute scale.

The present urgency is to begin thinking within the context of the whole planet, the integral Earth community with all its human and other-than-human components. When we discuss ethics we must understand it to mean the principles and values that govern that comprehensive community. Human ethics concerns the manner whereby we give expression at the rational level to the ordering principles of that larger community.

The ecological community is not subordinate to the human community. Nor is the ecological imperative derivative from human ethics. Rather, our human ethics are derivative from the ecological imperative. The basic ethical norm is the well-being of the comprehensive community and the attainment of human well-being within that community.

Here we find that we are dealing with a profound reversal in our perspective on ourselves and on the universe about us. This is not a change simply in some specific aspect of our ethical conduct. Nor is it merely a modification of our existing cultural context. What is demanded of us now is to change attitudes that are so deeply bound into our basic cultural patterns that they seem to us as an imperative of the very nature of our being, a dictate of our genetic coding as a species. This reference to our genetic constitution is indeed the issue.

Our genetic coding is more comprehensive than our cultural coding. Human genetic coding is integral with the whole complex of species codings whereby the Earth system remains coherent within itself and capable of continuing the evolutionary process. For a species to remain viable it must establish a niche that is beneficial both for itself and for the larger community. The species coding of the human carries within itself all those deeper physical and spiritual spontaneities that are consciously activated into cultural patterns by the genius of human intellect, imagination, and emotion. These cultural patterns are handed down as traditions, which form the substance of the initiation rituals, educational systems, and lifestyles of the various civilizations.

Our cultural traditions are constantly groping toward their appropriate realization within the context of an emerging universe. As things change, the traditions are forced into new expressions or into an impasse that demands a new beginning. The norm for radically restructuring our cultural codings forces us back to the more fundamental species coding, which ties us into the larger complex of Earth codings. In this larger context we find the imperative to make the basic changes now required of us. We cannot obliterate the continuities of history, nor can we move into the future without guidance from existing cultural forms. Yet, somehow we must reach even further back, to where our human genetic coding connects with the other species codings of the larger Earth community. Only then can we overcome the limitations of the anthropocentrism that binds us.

Perhaps a new revelatory experience is taking place, an experience wherein human consciousness awakens to the grandeur and sacred quality of the Earth process. Humanity has seldom participated in such a vision since shamanic times, but in such a renewal lies our hope for the future for ourselves and for the entire planet on which we live.

10

The NEW POLITICAL ALIGNMENT

THE OLDER TENSION IN HUMAN AFFAIRS BETWEEN CONSERVATIVE and liberal based on social orientation is being replaced with the tension between developers and ecologists based on orientation toward the natural world. This new tension is becoming the primary tension in human affairs.

So too the political tension between the empires and the colonies is being replaced by an economic tension between village peoples of the world with their organic modes of agriculture and the transnational corporations with their industrial agriculture.

This new alignment should not be taken as if the ecology movement were a New Left movement or a new liberalism. For the ecology movement has moved the entire basis of the division into a new context. It is no longer a division based on political party or social class or ethnic group. It is a division based on the human as one of the components within the larger community of the planet Earth.

In this new alignment those committed to industrial-commercial development of natural areas see this development as inherently progressive. Those committed to the integrity of the natural world and their indigenous peoples see this development as degradation, since the intrusion of the human into the life systems of the planet has already gone beyond any acceptable limits.

To the one group the human is considered primary in terms of reality and value while the larger, more integral Earth community is a secondary consideration. In the other group the integral Earth community (including the human) is seen as primary while human well-being in itself is seen as derivative. The one insists that the natural life systems must adapt primarily to human purposes. The other insists that the human must adapt to the priority of natural life systems. Ultimately there must be a mutual adaptation of the human and the natural life systems.

Reconciliation of these tensions is especially difficult because the commercial-industrial powers have so overwhelmed the natural world in these past two centuries that there is, to the ecologist, serious difficulty in further adaptation of natural systems to the human. Oppression of the natural world by the industrial powers has so interfered with the functioning of natural forces that we are already into an extensive disruption of the biosystems of the planet at the expense of the health and well-being of both humans and the natural world.

We cannot mediate the present situation as though there were some minimal balance already existing that could be slightly modified on both sides to bring into being a general balance. The violence already done to the Earth is on a scale beyond acceptability. It can only be considered as the consequence of a severe cultural disorientation. The change required by the ecologist is a drastic reduction in the plundering processes of the commercial-industrial economy. Until this is recognized there can be no way in which an acceptable reconciliation can be attained.

Yet we are so deeply committed to the exploitative mode of relating to the natural world that those in control of the great corporations can hardly think about modifying the exploitation in any significant manner. Even official movements toward "sustainable development" must be recognized as efforts to avoid the basic issue. Our sense of reality and of value has been so fully committed to the norms governed by the industrial process that such an abrupt shift is too difficult for serious consideration. These industrial norms of procedure are now functioning on a global basis through the transnational corporations.

These corporations, in alliance with the governments of the world, are now related to or organized into such establishments as the World Bank, the International Monetary Fund, the World Trade Organization, the International Chamber of Commerce, the World Business Council for Sustainable Development, and the International Organization for Standardization. Bonding of common interests has become so coordinated that it is increasingly difficult to escape not only their influence but their control over the various nations and cultures of the world.

So influential is the present commercial-industrial order that our dominant professions and institutions are functioning in this context; not merely our economic system, but government, jurisprudence, the medical profession, religion, and education. Every aspect of life has been absorbed into the commercial-industrial context. We seem not to know how to live in any other way. In the industrialized nations the automobile, the highways, parking lots, shopping malls, all seem to be necessary for survival at any acceptable level of human well-being.

Through the Internet a more extensive range of human transactions will be carried on without travel or physical presence, yet this will not remedy or remove the waste heaps, polluted waters, sterile and eroded soils, forests devastated by clear-cutting, toxic chemicals, radioactive waste, the thinning ozone layer. We see all this, yet we continue creating these chemicals, clear-cutting the forests, polluting the waters, piling up enormous waste heaps, destroying wetlands. We do this even though the industrial bubble is already dissolving. The end of the petroleum-based economy is in sight. Yet even now the commercial-industrial world insists that this is the only way to survive.

The tendency is to insist that ecologically oriented persons will accept the existing situation with some slight modifications. The system itself must continue in the existing pattern of its functioning. The alternative, the radical transformations suggested by the ecologists—organic farming, community-supported agriculture, solar-hydrogen energy system, redesign of our cities, elimination of the automobile in its present form, restoration of local village economies, education for a post-petroleum way of life, and a jurisprudence that recognizes the rights of natural modes of being—all these are too unsettling. Even though such books as Rachel Carson's *Silent Spring* are proving to be valid statements of the future that awaits us, they are still considered as too extreme to be accepted.

Never before has the human community been confronted with a situation that required such sudden and radical change in lifestyle under the threat of a comprehensive degradation of the planet and its major life systems. The difficulty can only increase. Tensions between capitalism and socialism, between liberalism and conservatism, are disputes over minor differences in comparison with the issues now before us. Both capitalist and socialist regimes are committed to ever-increasing commercial-industrial exploitation of the resources of the planet. Neither is acceptable to the ecologist.

Fixation on the primacy of industry in the well-being of the human is producing a recession of the basic resources of Earth which is now a permanent condition. This recession is not a temporary economic recession of any one nation, nor the recession of some financial or commercial arrangement, it is an irreversible recession of the planet itself in many of the most basic aspects of its functioning. The Earth simply cannot sustain the burden imposed upon it. The air in many places has become polluted. The water of the planet is toxic for an indefinite period of time. The soils of the Earth are saturated with chemicals. We have only the slightest idea of the consequences for the physical and psychic life of the human community, especially for the children who have lived in this chemically saturated environment since the day of their conception.

Physical degradation of the natural world is also the degradation of the interior world of the human. To cut the old-growth forests is not simply to destroy the last 5 percent of the primordial forests left in this country. It is to lose the wonder and majesty, the poetry, music, and spiritual exaltation evoked by such awesome experience of the deep mysteries of existence. It is a loss of soul even more than a loss of lumber or a loss of money. Loss of spiritual, imaginative, intellectual, or aesthetic experience is considered irrelevant by the developers as soon as a territory is identified as a place where money is to be made. In North America, even after taking 95 percent of these forests, developers insist on the right to cut the few timberlands that survive, while speaking of the extreme demands of the ecologists.

The severity of the tension between the developers and the ecologists can only be fully realized if, in addition to what has already been indicated, we understand that the exploiters have been in control of the North American continent since the beginning of its settlement by Europeans in the seventeenth century. Americans have never known any other way of life. The original settlers came here for religious freedoms but also for a "better" life than was available in the European world. The spaciousness of the continent, the luxuriance of its coastlands, its woodlands, its fertile soils, the beaver and deer and buffalo—all these seemed, in their abundance, to be beyond the capacity of any human force to diminish in any significant manner. The attrition of most life forms has been severe in these past few centuries.

Then came the capacity to exploit the coal deposits, the gold fields, the copper and iron ores; the skills to build the canals, the railroads, the highways; the ability to dam the rivers for irrigation and for power at a thousand different places. All this was done with a certain arrogance of the settlers, from the beginning. The rights of the indigenous peoples, the rights of living species, the rights of natural modes of being to exist, none of this evoked from the settlers any adequate sense of responsibility for their actions. When the chemical and electronics industries were established, when the power systems were put into place, when automobiles began to spread their exhaust over the countryside, even these events caused no adequate reflection or even interest in what was happening. Waste was simply poured into the air or dumped into the rivers or used as fill for wetlands.

Only the bright side of all this development was seen. The dark side, the toxic waste, was denied, ignored, hidden from sight, buried. Now, when the immense amount of such waste can no longer be hidden, when the poisons begin to affect the health of the populations, when the lead in the air and in the paints begins to affect the brain functions of children, when the "Love Canals" are identified, when the people of Louisiana begin to realize how extensively the countryside along the Mississippi has become saturated with chemicals, then the new alignment of forces begins to take shape.

The assault of developers on the ecologists has already increased in its pervasiveness and intensity. A person need only read *The War Against the Greens* by David Helvarg to understand the extent of this opposition. Insensitivity toward the devastation of the natural world led to disregard of the environmental issue throughout both the 1992 and the 1996 campaigns for the United States presidency. The most acute antagonisms of the past have seldom evoked such deep feelings of being threatened. Yet a polarity has evolved that now finds expression in every aspect of contemporary life, in our social and political and economic institutions, in our professions of medicine and law, in our educational programs, in our religious traditions. This polarity in life attitude pervades the public and private order of our society.

There will, naturally, be an infinite number of variations in the emphasis that will be given to various plans of action. But the main outlines of the tension are clearly evident. The tensions created will ultimately be even more severe than the capitalist-Communist tension that dominated political-social activities of the human community from the publication of *The Communist Manifesto* in 1848 until 1991, when the Soviet collapse occurred and left the capitalist world and its market economy in control.

In understanding these new tensions a person need only read a few surveys, such as the attack on the ecologists in *The True State of the*

Planet, a book edited in 1996 by Ronald Bailey, or *Dreams: Imperial Corporations and the New World Order,* which identifies the controlling power of the corporations, by Richard Barnett and John Cavanaugh. In addition to these a person might read *The Ultimate Resource* by Julian Simon, someone who argues that there is no real resource problem, population problem, or soil problem.

Yet there is still a tendency to think of ecologists as radical, romantic, or trivial New Age types. If by clear-cutting the last 5 percent of the surviving old-growth forests we provide jobs for the present, then clear-cutting is justified. This is the realist position. Forests are seen as so many board feet of lumber whose primary value is to be cut down for human use. The sense of meaning, of entry into the mysteries of existence, the grandeur experienced in their presence, all these are marginal to the essential thing of life, which is to exploit the forests for their passing human use and their monetary value.

Such issues require a reorientation of all the professions, especially the legal profession, which is still preoccupied with individual "human" rights, especially with the limitless freedom to acquire property and exploit the land. The number of lawyers hired by single corporations to defend themselves against any limitation of their perceived rights to exploit the natural world is evidence of the strange principles of jurisprudence that allow the devastation of the planet to proceed. Universities are still preparing students for professional careers in the industrial-commercial world even as this world continues its planetary destruction. The medical profession is only beginning to recognize that no amount of medical technology will enable us to have healthy humans on a sick planet.

A new awareness is emerging, however, throughout every realm of human activity. The term *sustainable development* is now the single most significant phrase in any discussion of these issues. This phrase obtained currency in the 1987 report *(Our Common Future)* of the World Commission on Environment and Development. It was later used to indicate the central concern of the United Nations Conference on Environment and Development held in Rio de Janeiro in 1992. So central is this phrase at the present time that it could be said that whoever owns it controls much of the discussion concerning the future.

Indeed, at the present time few persons would directly confront the proposal that development can no longer be as unlimited as it has been in the past. So prevalent is this phrase, sustainable development, and so widespread the claim to be acting with regard to the environment, that the deeper question has now become the question of authenticity in fulfilling its demands. Are contemporary commitments to safeguarding the environment merely up-front appearances with little substantial regard for the natural world, or is there a true commitment to limit industrial activity so that no real harm is done to the ecosystems of the planet?

The more realistic response to this phrase is that development is simply not sustainable. What is needed is a sustainable way of life. Paul Hawken goes further than sustainability with his proposal that a "restorative economy" is already in process. This view is presented in his book on *The Ecology of Commerce* (1993) and carried out in its basic principles through the movement known as "The Natural Step." Another more rigorous critique of the corporations is presented by David Korten, *When Corporations Rule the World* (1995). Both are working toward a depth understanding of the present situation with suggestions as to a viable way into the future.

David Korten makes proposals for the sequence of intermediate steps needed if we are to move into a sustainable mode of human presence on the planet in a later book, *The Post-Corporate World: Life After Capitalism.* A further observation might be made that a *sustainable* mode of survival at our present level of economic well-being in the industrialized countries is hardly possible as a universal attainment. It is estimated that to support our present Earth population at the level enjoyed in North America would require two or three planets.

The more ultimate question has to do with the "soul" of the future as this finds expression in the single life principle of the planet Earth. There is much consideration of the physical and biological modes of survival with relatively little comment on the soul of the future. Here we are mainly concerned with the "soul" as the shaping spirit within any vital process. These, the inner spirit and the outer form, are two distinctive aspects of a single mode of being. In considering the soul of the future, I am concerned with the inner vision that we need if we are to make the intellectual, social, economic, and religious adjustments required for a viable future.

That the human and other components of Earth form a single community of life, is the central issue of the Great Work. We can hardly repeat too often that every mode of being has inherent rights to their place in this community, rights that come by existence itself. The intimacy of humans with the other components of the planet is the fulfillment of each in the other and all within the single Earth community. It is a spiritual fulfillment as well as a mutual support. It is a commitment, not simply a way of survival. Anything less, to my mind, will not work. The difficulty we confront is too great. The future is too foreboding. We need to think of twice the present human population facing the future with half the resources. The next generations need a truly inspiring vision of the wonder and grandeur of life, along with the beginnings of the new technologies they will need.

The profoundly degraded ecological situation of the present reveals a deadening or paralysis of some parts of human intelligence and also a suppression of human sensitivities. That exploitation of the Earth is an economic loss should at least be evident, especially when we observe such extinctions as have occurred in the seas. There we can observe that some species of fish have become commercially extinct because humans would not limit their take to the reproduction rate of the fish, even though this reproduction rate was almost astronomical in the abundance of its production, as was the case with salmon in the Pacific and cod in the Atlantic.

When the proposal is made that we must continue what we are doing "in order to provide jobs" it must be considered as an unacceptable solution when a much greater abundance of jobs is available for repairing the already damaged environment. In all of these instances we can see a disposition toward biocide, the destruction of the life-systems of the planet, and geocide, the devastation of the planet itself, not only in its living creatures but in the integrity of the nonliving processes on which the living world depends.

Read the publications of the business world—*Fortune, The Economist,* or the *Wall Street Journal*—to observe the abandonment of any discipline that would limit the moneymaking concern of our industrial society, for it is precisely by this grasping after greater wealth to sustain a "better life" that we perceive "progress." The pathology of this attitude is the limitless straining after what cannot be attained by any level of consumerism. As with any addiction, the addiction itself is seen as the way to life. The authentic remedy, the only valid way to life, is perceived as too painful for acceptance.

What we propose here is not a solution of the issue but a clarification of the fact that the real issue before us is no longer finding expression in terms of liberal and conservative but rather in terms of the ecologist or environmentalist on the one hand and the commercial-industrial establishment on the other. A new alignment of forces is taking place throughout every institution and every profession in our society.

It is important to understand this new situation, the inherent difficulties of reconciliation, and the new language that has come into being. Only in this manner can we appreciate the true nature of the issues under discussion and the magnitude of change required in shaping a viable mode of human presence on the planet Earth for the future. All our professions and institutions need to be reinvented in this new context. We must in a manner reinvent the human itself as a mode of being. Eventually this implies rethinking the planet and our role within the planetary process.

11

The CORPORATION STORY

AMONG THE MORE SIGNIFICANT CONCERNS OF THIS TRANSITION period into the twenty-first century must be the modern industrial, commercial, and financial corporations. We need to understand the larger significance of these corporations in American society, in the human community, and in the functioning of the planet Earth.

The corporations, in their ambivalent commitment to financial profit while making progress in human affairs and providing comfort and security for people, are among the principal instruments for devastating the planet. There are other historical and cultural causes found deep in the course of Western civilization, yet these other causes have found their most effective expression in the structure and functioning of the modern corporation.

The term *corporation* is being used here in reference to the industrial, commercial, and financial corporations as these have existed in the last few centuries in the economic life of American society. These corporations are the organizing centers directing the discovery and use of modern science and technology in the quest for human benefit and financial gain by exploiting the living and nonliving resources of the planet.

As long as these corporations continue in their relentless exploitation of the planet through their oil wells, their automobile manufacturing, their chemical compounds, logging projects, road building, and their assault on the marine life of the seas, then the biosystems of the planet will continue to be extinguished. The entire range of life development for the past 65 million years will be threatened. Life will be unable to provide the high level of intellectual, imaginative, emotional, and spiritual fulfillment demanded by the very nature of our human mode of being.

Accomplishment of a program of integral survival of the planet, and of the human community, requires that the dominant profit motivation of the corporation endeavor be replaced with a dominant concern for the integral life community. To seek benefit for humans by devastating the planet is not an acceptable project. The ruin brought about on this planet over the past two centuries causes a certain foreboding concerning the possibility of the corporation, as we have known it in the past, reforming itself so that it will be a support rather than an obstacle in achieving a viable future. Yet this is the challenge that is before us. We will change or we will die in a major part of our inner being.

Although the corporations are constantly developing, absorbing and striving against one another, there is a unity to the corporation endeavor. The tension among the corporations drives each to greater intensity of activity as they rival, threaten, and support one another. This mutual support amid the stress of their diversity is what constitutes the "market economy." They are dependent on the same resource base, the same citizenry, the same media technologies. They are served by each other. They are committed to the same market economy. They have the same opposition to any national or international governmental regulation. They are especially resistant to any restraints on their activities based on protection of the environment.

The corporations have accomplished much by bringing about improvements in human well-being through the increase in available food, clothing, and housing, as well as making available mechanical energy for burdensome and repetitious labor, ease in travel and transportation, and increasing medical technologies. The corporations have alleviated human misery in many ways. Yet they have also caused vast areas of social disarray within the human community and the loss of genetic variety in the sources of our food supply.

A person needs only to read the writings of Vandana Shiva, the economist-social critic from India concerning the deleterious consequences of the so-called Green Revolution, a program that has been extensively praised for its accomplishments by increasing the food supply throughout the world. As we are told by Vandana Shiva, "Since 1970 the Green Revolution experiment has ruined land that could have produced much more food. A third of India has become wasteland. Half the Punjab, once known as India's wheat basket, now lies unproductive. Malnourishment haunts 60 percent of India's children" (quoted in Breton, p. 214).

We are beginning to ask about the real quality of life achieved, the environmental and social costs, also about the more lasting consequences of these so-called improvements in human life and in the integral functioning of the natural life systems of the planet. India, Indonesia, and the Philippines are three countries that need to be studied in this context.

There are several basic critiques to be made of the corporations as they exist in the United States. They have obtained the natural rights of individual citizens without assuming responsibility in proportion to their influence on public concerns. They have devastated the natural endowment of the North American continent. They have corrupted the government. They have relentlessly harassed the public through newspapers, mail, and magazines, through signs and billboards on the highways, through telephone and television, through sponsorship of sports and cultural events, through exploitation of the wonder of children, of the female form, of the sacred seasons of the year. They have even used the sky as a billboard for advertising. In the social order they have not given the working people their share of the profits earned through the effort of these same people.

Through all these impositions the corporations have taken possession of human consciousness in order to evoke the deepest of psychic compulsions toward limitless consumption. This invasion of human consciousness has brought about deleterious effects throughout the moral and cultural life of the society as well as the impoverishment of the Earth. Yet the corporations are so basic to contemporary life that a central purpose of contemporary education from high school through college, and even through professional training, is to prepare younger persons for jobs within the corporation context.

That the corporations are the dominant powers on the national, and now on the global scale, is clear. In the United States they have extensive controls over city, state, and federal governments and agencies. The less developed cities and states compete in offering benefits to the corporations to settle there in the form of tax abatements, road building, public utilities, and various easements in local ordinances. The less industrialized states offer extravagant incentives from public funds to private corporations willing to relocate in the region. This has become a common practice even in the more advanced states.

Even beyond their national context, the modern corporations in their transnational activities might now be considered as the most influential human institutions functioning on the planet Earth. While the state and federal governments have some control over the corporations through the charters granted and the regulations imposed, these corporations, by virtue of their status as citizens under the Constitution, have all the inherent freedoms and rights and privileges granted to individuals by the Constitution. In this manner they escape control over their activities and are responsible to no one except themselves and their stockholders. None of the political empires of past ages had anything like the control over land and peoples now held by the more powerful corporations of the twentieth century, nor has any economic system had such effective technologies for exploiting the resources of the planet. The great transnational corporations individually have assets far beyond the combined assets of half the nations of the world.

These corporations now own or control the natural resources of the entire planet directly or indirectly. They provide the jobs and pay the salaries of many peoples of the world. They make and sell the products. They set the prices. They extract the various ores from Earth, fashion and sell the products. Yet they have no proportionate responsibility for the public welfare. Indeed they insist on being recipients of government grants and exemptions—"corporate welfare," as it is now called.

This control by the corporations had its beginning in the period when the colonial powers of the European world assumed the right to invade, possess, and exploit the entire planet for the benefit of the religious, political, and economic powers then in control of the nations of Europe. Of these regions of the world occupied by European powers, the foremost in its influence over the modern world has been the central region of the North American continent in the territory occupied by the English beginning in 1606.

This region was settled by groups that were chartered by the king of England in the manner of corporations to occupy and develop a vaguely defined area for the primary benefit of England but also for whatever benefit colonists sought in their own ambitions. To understand the sequence of events leading from this early period to the present we will take a brief look at the history of the corporation in what became the United States.

The early phase of the corporation here began with the land corporations, such as the Virginia Company (1606) and the Plymouth Company, chartered in 1606 and then rechartered as the New England Company in 1620. Later settlements were made by proprietary grants made to individual persons, to William Penn in the region now known as Pennsylvania in 1681, to Lord Calvert in Maryland in

1632, and to the eight proprietors in the Carolinas in 1663. In the New York area extensive land grants were made to particular proprietors such as Rensselaer, founder of the Dutch West India Company. When he died in 1644 he owned property that now includes the complete counties of Albany, Columbia, and Rensselaer in the state of New York. In Virginia in the eighteenth century William Byrd II owned 100,000 acres. Robert Carter left an estate of 500,000 acres. George Washington and many others constituted a landed elite with possessions that foreshadowed the corporation wealth to be attained in later times. This class of elite proprietors was the forerunner of the corporation personnel that would later take control of the economic life of the continent.

In 1812, when the Land Grant Office was established, the federal government possessed 756 million acres of land. To get the land into the possession of the people and to finance government expenses, the land was offered to the public at minimal cost. This opportunity led to the rise of land speculators who bought large areas and later sold them for extensive profit. This attitude toward land as a commodity to be freely bought and sold for financial profit by individual entrepreneurs was a further development of the general attitude toward land that had been developing in Europe throughout previous centuries. Progressively, reverence for the sacred dimension of the natural world, even for the sense of land as a commons, which had so far survived deep in European consciousness, was further diminished.

The settlers found difficulty in relating to this continent in any creative manner. Some ancient fear of the wilderness in Western civilization led either to a direct assault on the various life-forms of the continent or to subjugation for some utilitarian purpose. Land was for settlement and possession. Soil was for cultivation. Forests were for timber. Rivers were for travel, for irrigation of the fields, and for power. Animals such as the wolf, the bear, and the snake were for killing. Animals such as the beaver, the deer, the rabbit, and the passenger pigeon were for the fur or the food they could provide. Fish, so abundant throughout the streams and rivers and along the shores, were for catching. North America was indeed a luxuriant continent awaiting human exploitation under the title of "progress" or "development."

There was a latent appreciation of the continent for its wonder and inspiration that would later appear in the naturalist writers, the scholars, the artists, the poets, and a few religious persons who could understand the need that humans had for the natural world to activate the inner life of imagination, emotion, and understanding and to convey a sense of the sacred. For the most part, however, the settlers brought their Bibles with them. That was all they needed for spiritual inspiration. They were hesitant to experience the deep spiritual communication provided by this continent. It might bring them to judgment for heresy.

There were persons with a scientific interest such as Cadwallader Colden (1688–1776), a physician who classified the plants in the New York region and who was in extensive communication with Linnaeus, the Swedish botanist who invented the scientific scheme of classification that is still used in botanical studies. Then there were other persons such as John Bartram (1699–1777), who collected plant species and founded a botanical garden in Philadelphia. His son, William Bartram (1739–1823), traveled through the southeastern states and collected both plant and animal species. John Lawson (d.1711) traveled through the Carolina region while taking extensive notes on the various forms of plant and animal life he discovered on the way. These were some of the persons who developed the scientific background for the later naturalists such as Henry Thoreau (1817–1862) and John Muir (1838–1914), who established a more intimate association with the natural life systems of the North American continent.

Such is the context for the later tension between the utilitarian approach to the natural world on which the later corporations based themselves and the more aesthetic and cultural intimacy with this continent that gave rise to the environmental movements throughout the nineteenth and twentieth centuries. Later, in the last half of the nineteenth century, there developed in this country a realization that some natural areas should be preserved, although this was, in the minds of such men as Gifford Pinchot (1865–1946), chief of the Forest Service (1898–1910), primarily to save areas for use by later generations. Pinchot's sense of the dominant utility value of the land caused the severe break in his relations with John Muir, whose appreciation of the land had a much more profound basis in the spiritual, aesthetic, and cultural needs of humans that the wilderness provided.

The second phase of the corporation in America might be identified as the canal and railroad phase. The canal phase began in 1817 when New York State decided to build a canal 363 miles long from the Hudson River to Lake Erie, a project that was completed in 1825. Such projects were carried out by corporations chartered by states and given rights of eminent domain, a right justified for the public benefit derived from it. In this manner the interior of the continent, with all its natural resources and with its vast possibilities for agricultural production, was joined to the city of New York, which then became the most advanced of the five leading commercial cities of the Atlantic coastal region: Boston, New York, Philadelphia, Baltimore, and Charleston. The early success of the Erie Canal led to the construction of numerous other canals joining the Atlantic side of the Appalachian Mountains to the interior of the continent, also joining such waterways as the Chesapeake and the Delaware. Farther in the interior, canals joined Lake Erie with the Wabash River, and Lake Michigan with the Illinois River.

Although they were owned by private corporations, the canals were subsidized by the state governments. Such financially profitable projects were developed with an enthusiasm that rose to almost a frenzy. Yet this period of canal-based commercial-economic expansion lasted, in its full intensity, for only some thirty years, until around 1850 when the railroads began to take over the transportation activity of the country.

The first railroads began operations in 1830 with the Baltimore and Ohio from Baltimore to Wheeling, the first commercial railway line in

America. This dominant canal-railroad period extends from 1817 until after the Civil War. The land grants made by the government to fund the railroad venture in the West was the vast sum of over 130 million acres. Such was the size of government grants to private corporations. Out of the immense grant to the Northern Pacific Railroad came the great timber corporations: Weyerhauser, Potlach, and Boise Cascade.

Later more extensive areas were made available in the western section of the country. The full story of the grants made to the Northern Pacific Railroad in 1864 is told in *Railroads and Clearcuts* by Derrick Jensen and George Draffan with John Osborn. They describe the deterioration of the biosystems of the region consequent to these land grants (p. 56).

While the first impressive fortune made in America was that of John Jacob Astor (1763–1848) through the American Fur Company founded in 1808, the fortunes made by railroading were even greater. These fortunes include those of Cornelius Vanderbilt, the New York Central; James Hill, the Northern Pacific; Edward Henry Harriman, the Union Pacific; Leland Stanford, the Central Pacific; Collis P. Huntington, the Southern Pacific; and a long list of others who founded and controlled the corporations that built the railroads throughout this continent. The facility in travel and transportation offered by the railroads made the entire continent available for intensive industrial and commercial development.

The third phase, the truly formative phase of the modern corporation, began after the Civil War with the ever-increasing discovery of oil and the invention of the electrical, the petrochemical, and automotive industries. The technological achievements of the post–Civil War period provided the energy and engineering skills needed for a new type of control over the continent. It was in this post–Civil War period that the United States became an urban industrial society dependent on wage earnings for fulfillment of the basic needs of personal and family life.

Two significant movements consequent to the corporation advance occurred at this time. The first was the early awareness of the harm being done to the natural world by the new industries and the beginning of the conservation movement. The other was the social protest against the exploitation of the working people who did the difficult and hazardous work in the mines, in construction jobs, in building and repairing the roads and bridges, in digging the sewers, firing the steel furnaces, and in the multitude of other exhausting jobs that men have been called to do in building our industrial world. It was a painful experience to realize that they were excluded from a proportionate share of the benefits derived from their work.

The inequality in the distribution of income from the various corporation ventures led almost immediately to a simmering resentment throughout the nineteenth century that has been referred to as "the other Civil War" by Howard Zinn in his *A People's History of the United States* (1980). These tensions between the workers and the industrial establishments culminated in the strike in Chicago at the McCormick Harvester plant in 1886 and the strike near Pittsburgh at the Homestead Steel Mill of Andrew Carnegie in 1892. At Homestead a number of workmen were killed and others were wounded by the Pinkerton Detective Agency hired to repress the strike.

The more significant writings in protest against what was happening in the industrial development of the country appeared at this time: Henry George's *Progress and Poverty*, in 1879; Henry Demarest Lloyd's *Wealth versus Commonwealth* in 1894. Andrew Carnegie defended the exaggerated accumulation of personal fortunes with his *Essay on Wealth* in 1889, an essay that came to be referred to as *The Gospel of Wealth.* Here he proposed that it was necessary for some persons to become extremely wealthy in order to carry out great enterprises on behalf of the community. His own personal sponsorship of cultural and educational projects beneficial to the people was evidence of the sincerity of his proposal.

The intense dedication of the American people to individualism, noted by Alexis de Tocqueville in *Democracy in America* (1836), when combined with intense attachment to private property was exactly suited to the evolutionary theory based on natural selection proposed by Charles Darwin in the *Origin of Species* (1859). This teaching popularized by Herbert Spencer in the phrase "the survival of the fittest," was advanced in America by William Graham Sumner (1840–1910) who began his teaching at Yale in 1872. Sumner communicated to his students a thorough conviction that progress in the human venture was brought about by the relentless struggle for existence.

In the view of Sumner, population increase, combined with the law of diminishing returns, constituted "the iron spur which has driven the race on to all which it has ever achieved. . . ." (Smith p. 141). The laws of evolutionary development were merciless. "The strong survived, the weak went down. Through this process of natural selection the race survived and improved. To aid the weak was to contaminate the gene pool and produce a negative effect on the evolutionary process. . . . The higher the organization of society, the more mischievous legislative regulation is sure to be" (Smith, p. 142). Against such teaching the Constitution was of little avail, and the Bill of Rights favored persons with extensive property possessions.

That the corporation could be so severe with its working people and so devastating to the natural world was not simply because of personal drive for power and possession but also because of a mythic sense that the industrial process under corporation control, driven by the doctrine of the survival of the fittest, was the predestined means for fulfillment of the historical destiny of humans. This destiny was the attainment of a kind of industrial-technological wonderworld, a state of profound human fulfillment, a vision of the future that appeared ever closer with the achievements of the new age of plastics, electronics, and computers that came after World War II. Such an achievement was considered ample justification for all the oppression to be imposed and all the devastation to be wrought on the way. The sense of progress as "control over nature" attained by human talents was manifested in economic competition in a realm of free enterprise. These two attitudes, derived from Darwinism, can be considered as the background of the industrial and corporate control of America. In their larger consequences these attitudes have led to the exploitation of the continent, and of the planet itself.

The entire American society was caught up in the transformation taking place. From its beginning until the present, the corporation has proclaimed that the public well-being could only be attained through a prosperous industrial, commercial, and financial establishment whose benefits were freely appropriated by the managerial and ownership class, with minimal payment to those who provided the labor and the skills needed for the process. Any government regulation of these establishments was considered an intrusion into the natural laws of the market economy governing the production and distribution of the goods.

Government, concerned with "establishing justice," insuring "domestic tranquillity," and "promoting the general welfare" as announced in the preamble to the Constitution, and the corporations, dedicated to limitless increase in personal profit, have seldom related to one another effectively except by sacrificing "establishing justice" and the "general welfare" for the appearances of "domestic tranquillity." When the government was challenged in its role of keeping public order in those years of struggle between the workers and the owners, the government consistently sided with the corporations in preserving the existing order against the workers whose energies were exploited.

Because the corporations had such power to resist any regulation by the government, they prospered through the legal and illegal use of public funds and public properties, such as the forests for logging, the rivers for damming, the mountains for mining, the grasslands for grazing. Much of this was through using influence on the legislatures of the country and by direct and indirect pressures exerted on the judiciary and the administration, largely through manipulation of the media.

Since the corporations controlled the instruments of production and since neither the states nor the federal government had any adequate legal structures to deal with the abuses of people, of public properties, or the despoliation of the natural environment, the corporations developed with little restraint through the nineteenth century and the greater part of the twentieth. Neither the Interstate Commerce Commission established in 1887 nor the Sherman Antitrust Act was able to prevent continued exploitation of the land. Even after the Environmental Protection Agency was founded in 1970, Congress was willing to support the Agency only in exceptional cases and only in a limited manner.

In *A History of American Law* by Lawrence Friedman we read that in the nineteenth century "the investment market was totally unregulated; no SEC kept it honest, and the level of promoter morality was painfully low. It was the age of vultures. In this period men like Vanderbilt, Jay Gould, and Jim Fisk fought tawdry battles over the stock market, the economy, the corpses of railroad corporations. The investing public was unmercifully fleeced" (Friedman, p. 513). It can be noted also that the Supreme Court formally recognized the corporation as a person before the law in 1886 in the case between Santa Clara and the Southern Pacific. Since this time corporation law has been among the most significant issues in transforming the mood and meaning of American law.

The corporation story of the twentieth century up to World War II is the story of continued expansion by appropriating the wealth of the continent and the labor of the people for private gain and limitless possessions. The biggest gains were through appropriation of public land and resources by mining, irrigation, grazing privileges, timber cutting, oil wells, and agricultural and transportation subsidies. Few industries in this country have come into being or prospered except through public subsidy. This is especially true of the great dams that came under corporate control and private benefit, built at public expense to provide energy, irrigation, and drinking water for the western part of the country. The story is thoroughly presented by Marc Reisner in *Cadillac Desert* (1986).

A fourth phase of the corporation began with the ending of World War II in 1945. Whatever the other consequences of the war, the extension of human concerns beyond national boundaries was among the most significant. The main instrument of this wider concern was the establishment of the United Nations in 1945 with its associated social and economic organizations and related institutions, the World Bank and the International Monetary Fund. In this manner the larger context was established in which human affairs have been conducted ever since.

Three terms have evolved in recent times to describe the work of the corporations in the United States: corporate libertarianism, corporate welfare, and corporate colonialism. These terms have all been dealt with in extensive detail by David Korten in *When Corporations Rule the World* (1994).

The term *corporate libertarianism* refers to the insistence of the corporation on the freedom to carry on its work as seems best in any manner not explicitly forbidden by law. This lack of any adequate controls over the corporations enables them to use the resources of the continent to extract their benefit and then to leave the waste-disposal problem as a public responsibility. Any pollution of the public domain is considered among the externalities, aspects of corporate functioning for which the public must bear the consequences on health as well as the financial costs. The agricultural corporations particularly show little concern for the disruption of the biosystems of the continent resulting from their use of chemical fertilizers, pesticides, and herbicides. When later they are required to absorb the expense of cleaning up the residue, they insist that this is a responsibility of the government, to be paid for by public funds derived from taxes.

The term *corporate welfare* means that the government best serves itself, the people, and the well-being of society by using public resources in support of the industrial or commercial corporations. There is to be minimal or no regulation by the government, even though almost every industry has come into being and survives with support from public lands and public funds. This is certainly true of the nuclear industry, also of the communications industries. These industries, along with such other industries as transportation, construction, and foreign commerce, were developed with heavy expenditure of public funds and appropriation of public lands. Much of this development, especially in the nuclear field, has been carried out by funds provided by the military with its almost limitless budget. Even the interstate highway system, begun in the 1950s in the regime of President Eisenhower, was 90 percent funded for defense purposes.

The present tendency, fostered by the corporation endeavor, is to see government as the source of financial funding. Yet as regards any controls or regulations government is seen as corrupting rather than protecting, as the enemy rather than the protector of the people, as an oppressive bureaucracy rather than as the primary source of public order and well-being in the community.

There seems to be little awareness that government, independent of corporation pressure, is the most powerful force the people have to offset the immense size of the corporations individually and in their combined influence over a nation's affairs. Big corporations require big government—unless the people are willing to accept the corporations as the government. This identification of the welfare of the corporation with the welfare of the people and of the government as the guardian of the people leads to absolute insistence that all benefits of the society be given first to the corporations. Yet their first obligation is considered to be not to the community but to their stockholders.

Now a new phase in the economy of the nations is developing, the national economy is becoming inadequate. All effective elements of the production or consumption of goods are passing into a global economy. One of its main achievements in the period after World War II had been to return political independence to the various peoples of the world in countries that had been taken over as political colonies by the more industrial countries of the world. Now, however, through new free trade policies the former political colonies have become economic colonies controlled largely by the transnational corporations through the World Bank and the International Monetary Fund. This new situation is designated as *corporate colonialism.*

Somewhat earlier in Western history the supreme human institution was the nation-state. This was the mystical reality of the nineteenth century, the ultimate referent in human affairs, the sovereign authority beyond which there was no appeal to any earthly power. This is still the accepted definition of the "nation," although the newly developed interdependence of nations has profoundly compromised the reality of absolute national sovereignty. Yet beyond the question of political interdependence is the reality that the corporations have now become the primary power centers both within nations and in the larger community of nations.

The nation-states have become subservient to the economic corporations. The corporations now function on a scale beyond any national boundaries. They have drawn the entire human community and the entire Earth into their control. The globalization of the human project as well as the globalization of the Earth economy is now reaching limits that will define the future in a new and decisive manner, for beyond the Earth no further expansion is possible in any effective manner.

In most discussion of the corporate world and its consequences, the main concern has been on social issues. Until recently, little consideration was given to the disastrous effects that the advancing industrial, commercial, and financial establishments were having on the life-systems of the planet.

This situation leads us to our present situation, the fifth phase of corporation history. This phase can be identified as the period of transition from a devastating phase of corporation economics to the period when the corporations recognize that a human economy can only exist as a subsystem of the Earth economy. While any comprehensive change just now would be beyond expectations, the existing corporations are finally beginning to recognize that they can only survive within the limited resources of the natural world.

The first occasion when some public recognition was given of the coming impasse between human demands and Earth resources occurred at the Stockholm Conference of the United Nations in 1972. After this conference the nations' representatives returned home and set up the first environmental protection agencies, with some reluctance on the part of the corporations. Then in 1980 the United Nations Charter for Nature was proposed by Zaire. After discussions in 1981 the Charter was passed in 1982. This could be considered as the finest official statement of human-Earth relations that has yet been given sanction by the world community.

Later, in 1983, the United Nations appointed a World Commission on Environment and Development. A report from that commission was submitted by its chairperson Gro Harlem Brundtland in 1987 under the title *Our Common Future.* Its basic message was that the human community must enter the future united among themselves and with careful consideration of the economic limitations of the Earth.

The enduring phrase that emerged from this report was that of sustainable development. This phrase became the subject for the United Nations Conference on Environment and Development (UNCED) held in 1992 where, in Agenda 21, an elaborate agenda was worked out in support of this project, an agenda that accomplished little in the realities of the corporation. While the phrase sustainable development is something of a contradiction, it has been modified into "sustainable future," a more acceptable phrase.

The difficulty is that until now the corporations have remained unconvinced of the need to align their own functioning and the limits of their activities to the possibilities of the Earth. Yet even now there is something of a growing commitment to sustainable development. The issue is there for the enduring future. It will not go away. Inevitably it will be modified into the hope for a viable future, even though CEOs and other corporate officials still insist that they know how to direct the destinies of the planet and its human community better than the natural forces that have until now controlled the functioning and the destiny of both the planet and its human presence. They insist in a special manner that only large-scale programs such as the World Bank, the International Monetary Fund, and the World Trade Organization can carry this out. They suggest this must be done with the support of the International Chamber of Commerce, the World Business Council for Sustainable Development, the International Organization for Standardization, and the World Industry Council for the Environment.

Opposed to this view are a multitude of competent persons committed to the view of the British economist E. F. Schumacher that "Small is beautiful," meaning that the deepest answers and the most viable economic programs into the future are those that have an intimate relation with the land. Only among a limited range of persons can true intimacy exist. So too only in limited sections of land can abundant and sustained cultivation be maintained. As indicated by David Ehrenfeld, the conservation biologist at Rutgers University, we have lost immense areas of intimate knowledge carried in traditional craft and in farming skills, knowledge that provides a relationship between the human community and the natural world that is immensely more bountiful and less destructive than that of large-scale business projects. Mass production and distribution is more expensive and less productive than we have recognized, especially in the field of agriculture.

What can be said of this fifth phase of the corporation in America is that since 1972 the issue has been progressively on the mind of the most significant public and private persons throughout the human community as the stark reality that must be dealt with in this coming century. Just how it evolves will determine the larger destinies of the human community as we venture into the twenty-first century.

When we hear corporations speak of "feeding the world" as a global enterprise, we can only reflect that feeding themselves belongs to each local community. It belongs to any people to be intimately related to the region of their dwelling. This includes a mutual nourishment. The land and all its living components nourish each other under the all-sustaining guidance of natural forces that bring us together, sustain us in being, and guide us to fulfillment of our diverse roles in the larger pattern of the planet on which we live.

As we reflect on this imposition of immense global corporations trying to take over responsibility for "feeding the world," we can only wonder at the reduction of the peoples of Earth to a condition of being nurse-maided by some few corporation enterprises. We might conclude that Mother Monsanto with her sterile seeds wishes to take over the role of Mother Nature herself. The people of the world need the assistance of each other, but only such assistance that enables them to fulfill their own responsibility for doing the essential things themselves. Village peoples everywhere, indeed all of us, need assistance within the pattern of our own inventive genius, not being reduced to a franchise of some distant corporation.

12

The EXTRACTIVE ECONOMY

WHEN WE CONSIDER WHERE WE ARE IN THE COURSE OF OUR historical destiny we might observe three events that could be considered as defining moments leading to our situation at the end of the twentieth century.

The first of these events occurred when the biblical-Christian emphasis on the spirituality of the human joined with the traditions of Greek humanism to create an anthropocentric view of the universe. At this time a certain discontinuity became possible between the human and the nonhuman components of the planet Earth. Throughout the earlier centuries of Christianity the integral relation of the human with the other components of the natural world was preserved. The natural world was considered as manifestation of the divine and the locus for the meeting of the divine and the human. Yet the sense of the Earth as a single integral community, with every being having inherent value and corresponding rights according to its mode of being, and with the human as one of the component members of this great community, was in the course of centuries diminished by excessive emphasis on the human as a spiritual being aloof from the physical universe.

A second historical moment occurred when this spiritual and humanist alienation was deepened into a feeling that the natural world was an actual threat to both the physical and spiritual well-being of the human. This feeling arose when the Black Death occurred in Europe from 1347–1349, a period when at least a third of the human population of Europe died. An even greater proportion of community leaders died. In the summer of 1348 fewer than 45,000 persons survived in Florence, Italy, out of 90,000 who were there at the beginning of the year. In Sienna, of 42,000 inhabitants only 15,000 survived (Meiss, p. 65).

Since the peoples of Europe knew nothing of germs, they had no way of understanding what was happening. They could only conclude that humans had become so depraved that God was punishing the world. The best thing to do was to intensify devotion and seek redemption out of the world. Along with a certain amount of moral dissolution caused by desperation a new intensity of spiritual dedication developed. A spiritual suspicion toward the world developed that has continued throughout the intervening centuries until our own times.

The division between the secular and the spiritual was intensified. Each was disengaged from and in a sense abandoned by the other. Thus, disengaged from any restraint imposed by spiritual concerns, industry and commerce, with the assistance of science and technology, seized control of the natural world and forced it, with great profit to themselves, into subservience to human convenience. Once we understand this background of our present situation we can proceed with our understanding of just what has happened. We might also recognize, with some clarity, just how we should proceed to bring about a remedy.

A third historical moment occurred in the last two decades of the nineteenth century when an even more severe situation arose. These were the critical years. In some sense the modern destiny of America, of the human community, and the planet Earth was determined at this time. These were the years of transition from an organic economy to an extractive economy. Modern technologies and the industrial establishment under the control of the modern corporation seemed to have effected an unqualified human conquest of the forces of nature. Indeed, they had achieved a control over nature never known previously in human history. The integral functioning of the geobiological systems of Earth that had governed the functioning of the planet for some 4.6 billion years came under the assault of humans determined to use the resources of Earth and the infinitely subtle functioning of nature in a manner immediately beneficial for humans, without regard for the consequences for the natural systems of the planet.

Only now do we understand that this was the time when we set forces into motion that would disturb the chemical composition of the air, water, and soil to an extent that would affect the entire network of organic life on the planet. The ozone layer protecting the organic forms of Earth from the ultraviolet rays of the sun would be weakened. The tropical rain forests would become largely destroyed. The residue of burning fossil fuels would lead to an excessive amount of carbon dioxide in the atmosphere and possibly cause a change in the global climate. Some of our most competent biologists in their knowledge of the biosystems of the planet would tell us that nothing at this level of species extinction has occurred in the 65 million years from the end of the Mesozoic Era until the present.

Little attention has been given to the consequences of basing the entire functioning of the human community on an extractive economy. An organic economy is by its very nature an ever-renewing economy. An extractive economy is by its nature a terminal economy. It is also a biologically disruptive economy. As long as we lived within the bounty of the seasonal renewing productions of the biosystems of the planet we could, apparently, continue into the indefinite future. But as soon as we established a way of life dependent on extracting nonrenewing substances from Earth, then we could survive only so long as these endured; or so long as the organic functioning of the planet was not overwhelmed by the violent intrusion involved in extracting and transforming these substances. A further difficulty came about from the contaminants that resulted, especially from the chemical industry.

Metals come generally in the form of ores which must undergo smelting processes. These processes give off pollutants which foul the environment. Liquids such as petroleum must be extensively transformed in order to supply fuel to be used by trucks, automobiles, aircraft, and generators, or by other machines capable of internal combustion. Petroleum can be used for the creation of fibers for the making of fabrics, also for making plastics. In all these cases there are toxic residues for which there are, at present, no adequate methods of disposal. These contaminants are dispersed into the air, water, and soil, resulting in the dispersion of toxic residue throughout the planet. Even as regards the benefits from petroleum, little attention is given to the fact that within another forty years, we will have consumed over 80 percent of the available supplies. By the close of the twenty-first century we will see the absolute end of petroleum as we have known it. The conditions of its production will never occur again.

Another difficulty with an extractive economy has been its use of engineering technologies to turn renewing resources into nonrenewing resources. This occurs when we exploit the soils of the Earth through chemical fertilizers, pesticides, or herbicides, to such a degree that we exhaust these soils or make them toxic. Even in the fishing industry, through our electronic instruments, our drift nets, and use of factory fishing vessels, we are able to exhaust the marine resources of the seas and rivers of the world in a manner that terminates their capacity for self-renewal.

Transition into an extractive economy had been long in preparation. It required the earlier centuries of dedication to useful knowledge as the only worthwhile knowledge. While a humanistic education based on the classical traditions of literature and thought was the main effort in education throughout the earlier period of American history, the quest for useful knowledge through the sciences gradually inserted itself into the curriculum from the high school years on through the college years and the years of professional training in graduate school. Useful knowledge as the ideal of education goes back to the purposes of the American Philosophical Society when it was founded under the leadership of Benjamin Franklin in 1754. Useful knowledge was thought of as knowledge that enabled humans to use the normal processes of nature to procure special benefits for humans.

A psychic compulsion developed, even perhaps a kind of mysticism of progress, that drove the commercial and industrial entrepreneurs as well as the scientists and engineers in their work. Some scientists indeed were attracted by the quest for knowledge. An immense number, however, were driven by the quest for control over the awesome powers in the world about them. Suddenly commercial entrepreneurs became aware of the possibility of financial profits to be made by using these powers to release the human community from age-old afflictions; this awareness combined with the attractions of a life filled with an abundance of Earth's delights. These were attractions not to be refused.

Many of these benefits were now available on a widespread scale if only we saw the Earth as a collection of natural resources rather than as a mystical entity to be revered, or as that larger community in which humans found the fullness of their life experience. The final need was for a properly organized human effort to assemble all the human abilities required to engineer this new mode of functioning to the planet.

The establishment needed to coordinate this entire process came into being at this time in the form of the modern corporation, financed by an abundance of capital, guided and controlled by boards of directors, and led by chief executive officers convinced that they could reorganize the planet in a much better way than nature was capable of doing. Their mission was to exploit this continent and then to exploit the planet as extensively as could be achieved within the limits of the consumer economy. The phrase used for this compulsion in America toward exploitation of the Earth through territorial acquisition and exploitation was Manifest Destiny.

Until the end of the nineteenth century there was neither the learning nor the scientific technologies to move the entire economic structure of the country out of its rural agricultural basis into an urban industrial setting. Only in the 1880s and the 1890s did the awesome powers of science find expression in the realm of mechanical, electrical, and chemical technologies. A vast complex of engineering endeavors was required, guided by extensive scientific research and inventive imagination.

The Bessemer converter for making steel had been invented in 1856. The high-speed internal combustion engine was invented in 1883. Methods for creation and control of electricity were discovered. Most significant was Thomas Edison's invention in 1876 of the first research laboratory dedicated to producing new inventions through scientific observation and experiment. The most spectacular of his early achievements was the invention of the incandescent electric light bulb in 1879. In 1880 the first electric generating station was built in London to power street lighting. In 1881 a similar generating station was built in New York.

When we look back on this period at the end of the nineteenth century, we see how many of today's dominant corporations in the oil, electricity, telephone, steel, chemical, pharmaceutical, paper, and automobile industries came into being then and in the decades immediately following. Because no other event in human history bears any proportionate consequence in the disruption of the planetary process, it might be helpful to list a few of these corporations and observe how many of these still have control over both human affairs and the planetary process.

In oil: Standard Oil Company of Ohio was formed in 1870, Atlantic Richfield in 1870, Exxon (as Standard Oil of New Jersey) in 1882, Mobil in 1882, Amoco in 1889, Royal Dutch in 1890, Texaco in 1902.

In utilities: American Telephone and Telegraph was incorporated in 1885, Westinghouse in 1886, General Electric in 1892.

In chemicals: Dupont, begun in 1802 was later incorporated in New Jersey in 1903, then in Delaware in 1915. Dow Chemical was incorporated in 1897, Abbott Laboratories in 1914. It was only when the German patents were seized after World War I that the chemical industry rose to its full expansion in America. Union Carbide was founded in 1917, Allied Chemical in 1920.

In pharmaceuticals: Johnson & Johnson was incorporated in 1887. Clinton Pharmaceutical Company, formed in 1887, became Bristol Myers in 1900.

In paper and lumber: Scott Paper Company was incorporated in 1879, International Paper in 1898, Weyerhauser in 1900.

In steel: Carnegie Steel was incorporated in 1873, Jones and Loughlin in 1889, U.S. Steel in 1901.

In automobiles: Ford Motor was incorporated in 1903, General Motors in 1908.

These corporations, most of which still survive after a century of dominant influence, took control not only of the human community but in some sense intruded into the functioning of Earth itself. In the human community they established themselves first in the economic realm, then in the political and educational realms. These and the other corporations founded in the European and Asian worlds have taken control of the entire human community and are even in possession of vast areas of the natural world.

The entrepreneurs directing the corporations became the dominant powers in all those basic establishments that have directed the course of human affairs through the twentieth century. The university establishments in their scientific research projects became increasingly dedicated to the service of the corporation: in its schools of business administration, its law schools, its training in communication skills, in its schools of political sciences. In the post–World War II period, the power of the corporations can be seen especially in the chemical and communications industries and in space exploration.

In the chemical industry, the great expansion occurred after World War II when new possibilities for the chemical industry were seen. In 1940 in America we were making half a million tons of industrial chemicals each year. In the 1990s we were, according to the latest *Norton History of Chemistry,* making 200 million tons of industrial chemicals each year (p. 654). These are chemicals that nature cannot, for the most part, absorb into its normal organic processes.

An indispensable factor in this entire sequence has been the legal profession and the judiciary. From the beginning of the nineteenth century the legal profession and the judiciary in America had bonded with the entrepreneurs and their commercial ventures, even of this early period, against the ordinary citizen, the workers, and the farmers. As Morton Horowitz, who occupies the chair of History of Law at Harvard University, tells us in his study, *The Transformation of American Law: 1780–1860,* "By the middle of the nineteenth century the legal system had been reshaped to the advantage of men of commerce and industry at the expense of farmers, workers, consumers, and other less powerful groups within the society" (Horowitz, pp. 253, 254).

The well-being of the commercial-industrial enterprise came to be identified with the prosperity of the nation in all its aspects as well as the well-being of the people. This identity of the corporate good with the welfare of the people has become so extreme that governments have come to consider their primary obligation as the support, without any effective control over the manner of their procedures, of the industrial, commercial, and financial corporations. Any effort of governments to regulate industrial, commercial, or financial establishments was considered interference with the proper functioning of the economic order. Governments had a primary purpose to subsidize, with public funds and resources, the predation on the natural world by the private corporations.

With the intellectual tradition of useful knowledge; with the scientific tradition and the research laboratory all dedicated to discovering new ways of human dominance over the natural world; with the rise of technological schools of engineering; with schools of economics and business administration preparing a multitude of students for positions within the corporation structure; with the news media, newspapers and periodicals, radio and television, billboards and sports events, all supported by and subservient to the commercial-industrial corporations; with the organizing of the legal profession and the structuring of government in support of the commercial-industrial enterprise; with the distant unconcern of religion; with spiritualities of withdrawal from the distractions of the marketplace; with the alliance of all these forces and conditions, it is little wonder that the natural world has been assaulted in an unrestrained fashion throughout the twentieth century.

There was only one further step to take in the mid-twentieth century: the move from the many national economies to a comprehensive global economy, an economy in which the corporations would be exempt from control by any government and not be bound by allegiance to any government in the world. Transnational organizations would function without restraint and with accountability only to themselves. The idea of being truly global corporations, which might have been thought unreal in an earlier period of nationalist economic development, became a kind of urgency during the last phases of World War II. At the 1944 Bretton Woods Conference, an economic organization of the entire planet was envisaged by the creation of a World Bank and an International Monetary Fund.

Through these organizations the less industrialized nations could be brought into the orbit of the industrialized nations, ostensibly for the benefit of everyone but effectively for the benefit of the existing financial powers. The less industrialized nations especially would be encouraged to borrow money for development of their natural resources to be shipped overseas to the industrialized nations. These nations would then be able to process these materials with extensive profit. In this manner the third world would enter into the world economy.

Even though the original intentions may have been to benefit all nations, such organization of the world economy would bring the less-industrialized nations into a global economic arrangement dominated by the more industrialized nations. Inadequate attention was given to the effect these arrangements would have on indigenous peoples throughout the world. Traditional cultures would come under new forms of stress, their agricultural economies would be disrupted, their resources would be plundered. Meanwhile they would end up with extensive debts they could never repay. They were, indeed, unskilled in the functioning of a money-based exploitative economy.

Later these transnational corporations would press on to establish a borderless world through a General Agreement on Tariffs and Trade (1967) to be accepted by the nations. This agreement would transform into the World Trade Organization (1995), an arrangement whereby the corporations could exploit the Earth freely with responsibility to no nation but with extensive power over the entire community of nations. The World Trade Organization, the World Bank, and the International Monetary Fund would, in turn, come under the influence of such organizations as the World Business Council for Sustainable Development (1995), an organization originally composed of forty-eight transnational corporations to develop a global economy under their overall direction, supposedly with all peoples benefiting but in fact with themselves the chief beneficiaries.

Among the ostensible purposes of the new economic organizations is the fostering of support throughout the business community for sustainable relations between the human economy and the geobiological functioning of the planet. Any reforms can only be apparent since we would still live in a world invented and sustained by the industrial way of life with its vast array of technological controls over the natural world. The planet and all human affairs would still be governed by a sense of the Earth as commodity, only we would supposedly be more cautious in our manner of using the Earth and its resources.

Since the dominant corporations would exist beyond the control of any nation, they would be self-referent in their structure, their functioning, and in their objectives. The good of the corporation becomes identified with the well-being of the community since it provides the jobs, pays the salaries, and produces the goods that people buy with their salaries. A kind of global company village comes into being. The political order becomes a subsidiary function of the corporation.

It is difficult to realize the dimensions of the consequent control over our lives. The corporations control our minds through their ownership or influence over the public media. They dominate governments by their financial support of selected candidates for political office and by the constant pressures they exert on legislation through lobbying. In this manner they oppose legislation to restrict corporations and support legislation that provides subsidies for corporations, funds now referred to as "corporate welfare." The extent of such corporate welfare, on a global scale, is sometimes estimated at well over $100 billion annually.

These activities lead to an attitude among the people that corporations are their protection and government their enemy. A so-called democratic-based market economy controlled by corporations with bases of operations in most of the nations of the world is presented as the new way to peace, prosperity, and all good things. Education of the young is largely to prepare students for their role in the world of industry, commerce, and communication. The private universities especially and many of the research institutes in state universities are supported by money that comes ultimately from the corporations for purposes that eventually enhance their own status.

We are into a new historical situation. The forces that we are concerned with have control not simply over the human component of the planet but over the planet itself, considered as an assemblage of natural resources available to whatever human establishment proves itself capable of possession and exploitation. The intellectual, cultural, and moral conditions sanctioning this process have already been worked out. The truly remarkable aspect of all this is that what is happening is not being done in violation of anything in Western cultural commitments, but in fulfillment of those commitments as they are now understood. Thus any critique or quest for betterment cannot be supported simply on the claim that the present situation is in violation of Western cultural or moral commitments. Our Western culture long ago abandoned its integral relation with the planet on which we live.

Obviously the universe, the solar system, and the planet Earth are the primary given in any discussion of human affairs. We awaken to a universe. We have no immediate access to anything intellectually or physically prior to or beyond the universe. The universe is its own evidence since there is no further context in the phenomenal world for explaining the universe. Every particular mode of being in the phenomenal order is universe-referent. While this was clearly understood by former peoples, it is also evident in our times, since our scientific endeavors do not provide us with any explanation of the universe except the universe itself. This acceptance by science merely confirms what we know from simple observation. The universe, in the phenomenal world, is the primary value, the primary source of existence, the primary destiny of whatever exists.

A second observation to be made concerning the universe is that it does not exist as a vastly extended sameness. The universe exists in highly differentiated forms of expression. So too the planet Earth exists as a highly differentiated complex of life systems. The only security of any life expression on Earth is in the diversity of the comprehensive community of life. As soon as diversity diminishes then security for each life-form is weakened. This has become abundantly clear in recent studies of the biosystems of Earth, such as *Biodiversity,* edited by E. O. Wilson in 1988, and Wilson's own book, *The Diversity of Life,* in 1992.

A third observation might be that these various forms of expression are so intimately related that nothing is itself without everything else. Nothing exists in isolation. Any being can benefit only if the larger context of its existence benefits. This law can be seen in the honeybee and the flower. Both benefit when the bee comes to drink the nectar of the flower: the flower is fertilized, the bee obtains what it needs for making its honey. The tree is nourished by the soil; the tree nourishes the soil with its leaves. It is the ancient law of reciprocity. Whoever receives must also give.

These aspects of the universe constitute what I would refer to as the ontological covenant of the universe. I would also note that the planet Earth fulfills this covenant with special brilliance of expression by the manner in which its hundred-some elements are shaped into the five spheres of which it is composed; namely, the landsphere, watersphere, airsphere, lifesphere, and the mindsphere. Each of these is further differentiated into the innumerable forms in which each finds expression. The wonder, of course, is the bonding into a single community of existence.

Especially in the realm of living beings there is an absolute interdependence. No living being nourishes itself. There exists a sequence of dissolution and renewal, a death-life sequence that has continued on Earth for some billions of years. This capacity for self-renewal through seeds that bond one generation of life to a successor generation is especially precious to the animal world, which feeds on the excess of plant life produced each year.

Every animal form depends ultimately on plant forms that alone can transform the energy of the sun and the minerals of Earth into the living substance needed for life nourishment of the entire animal world, including the human community. The well-being of the soil and the plants growing there must be a primary concern for humans. To disrupt this process is to break the Covenant of the Earth and to imperil life.

Disruption of the biological integrity of the planet is the indictment that must be brought against the extractive economy. Only a restoration of the biological integrity of the planet within its various bioregions can assure the integral survival of Earth into the future. Our primary concern must be to restore the organic economy of the entire planet. This means to foster the entire range of life-systems of the planet. All are needed. It means also that we must establish our basic source of food and energy in the sun, which supplies the energy for the transformation of inanimate matter into living substance capable of nourishing the larger biosystems of Earth.

Among the primary evils of contemporary industry is that it is founded on uniform, standardized processes. This is especially devastating in agribusiness, which demands uniformity in its products. Nature abhors uniformity. Nature not only produces species diversity but also individual diversity. Nature produces individuals. No two days are the same, no two snowflakes, no two flowers, trees, or any other of the infinite number of life-forms. Since monoculture and standardization are violations of both the universe covenant and the Earth covenant, we need to foster a new sense of the organic world over the merely mechanical world.

Our concern for space exploration, in the expectation that we will have used up Earth and will need to move the human venture out onto other planets, is to waste irreplaceable resources and to neglect much-needed research into the organic world of this planet. Our excitement over the possibility of colonizing Mars is something of a childlike delight. We imagine something strange and exciting in some faraway place while we remain insufficiently interested in the wonders in our immediate surroundings and their well-being in the future. It's like the ancient fable by Aesop about the dog with the bone seeing his image in the water, then diving in to get the bone imaged there—a fable reenacted in our times with similar consequences.

Even as regards this planet we need to esteem this planet and its functioning in the depths of their mystery. The greatest of human discoveries in the future will be the discovery of human intimacy with all those other modes of being that live with us on this planet, inspire our art and literature, reveal that numinous world whence all things come into being, and with which we exchange the very substance of life.

13

The PETROLEUM INTERVAL

THE STORY OF THE LATE NINETEENTH AND THE ENTIRE twentieth century has been largely the story of petroleum, its discovery and use by humans, and the social and cultural consequences in human society. The story of the twenty-first century will be the story of the terminal phase of petroleum and the invention of new patterns of human living in relation to Earth's resources in the post-petroleum period. From its beginning in the mid-nineteenth century to its terminal phase, the role of petroleum in human affairs will have lasted two hundred years, from the mid-nineteenth until the mid-twenty-first century. These years, the glory years of the industrial period and the devastating years of the Earth, might be designated as the Petroleum Interval.

The consequences of these years in extinct species, toxic residue, and disturbed ecosystems will remain into the indefinite future. The consequences of scientific discoveries, technological skills, and new energy technologies in health benefits originating in this period will also remain. But even as we recall these benefits we must also be conscious that the story is still in process. Inventions that we thought were pure benefits in earlier days come with associated difficulties much greater than we realized. Our hybrid grains, our irrigation projects, our automobiles—all these have negative consequences we seem never to have suspected. Our antibiotics and our sprays against insects are, in many instances, evoking even stronger germs and more vigorous insects.

As we seek the far-reaching adjustments needed for a more viable way of life, we are finding that we are now so conditioned by our dependence on petroleum and its benefits that we can hardly imagine life without these benefits. To discover how we will move from a non-sustainable petroleum-based economy into some alternative form of sustainable economy is the problem. Just now, in this transition period into the twenty-first century, no comprehensive program seems to be available. Our efforts in every field of human activity, in economics, social structures, legal enactments, education, scientific research, in spiritual and religious life all need to be directed toward this restructuring of human life in a more integral relationship with the planet. This relationship will enable us to survive in a state of well-being in the post-petroleum period.

We might expect that the twenty-first century will also be the period when we will recover many of the valuable insights and skills in the art of living that were lost during the period when we were fascinated with new areas of knowledge. The craft skills associated with village life need to be recovered. A good example of achievement in this regard can be found in the village of Gaviotas, east of the Andes in Colombia. With the assistance of trained engineers and technical experts, the founder of the village, Paolo Lugari, has led the village in devising an energy system using solar and water power. He has also developed an agriculture program suited to the soil, a forestry program, a way of increasing the fish in the rivers, and a way of feeding the cattle on specially grown grasses. The people of Gaviotas believe their achievement can be adapted for third world people everywhere.

They have attained a remarkable self-sufficiency within their local situation (Weisman, 1998). Bill McKibben in *Hope, Human and Wild* (1995) tells the story of Kerala, a state on the southwestern coast of India, where in the midst of poverty and minimal resources the people have developed a way of life that in its sense of human fulfillment matches in many ways and may even be more humanly satisfying than life in more affluent circumstances.

But while we begin our programs for renewal in a post-petroleum period in America, we need to understand its historical influence over our lives and the difficulties we must deal with as the petroleum diminishes and then is gone forever. Much of what has been happening here and how we might respond to it in a post-petroleum period was outlined some years ago by Amory Lovins in his presentation of *Soft Energy Paths* in 1977.

But first the story itself. Petroleum has been used for thousands of years, not only in the classical civilizations throughout the Eurasian and North African worlds, but also by the indigenous peoples of the world. Although petroleum is for the most part contained in vast pools held within the rock formations deep within the Earth, it has built up such pressures that it seeps through to the surface, where it has been available for a variety of purposes: to caulk floating vessels for travel over water, for sealing baskets to carry liquids, for healing injuries.

The sheer volume of petroleum contained in the Earth is amazing. Estimates of the amount yet to be discovered are constantly found to be too low, even in the later years of the twentieth century. So too the places where petroleum is located are more widespread than we realized. We find it in North Africa and Nigeria, in the countries along the Persian Gulf, in Central Eurasia, in Indonesia, in South America, along the shores of Colombia and Venezuela, in the Gulf region off Mexico, in the Southwest and in the offshore Pacific region of the United States, in Alaska, in Canada, in the North Sea.

Yet, although the volume of petroleum present on the planet is immense, it is ultimately limited. From the extravagant rate of its use in the late twentieth century it seems that few people in positions of economic or political power have been seriously concerned with the problem of adjustment to the period when the supply of petroleum will be first diminished and then exhausted. Now, at the beginning of the twenty-first century, there is widespread agreement that by the mid-century some 80 percent of the petroleum on the planet will be exhausted, leaving a planet unable to support the way of life developed during its period of abundance. Such is the conclusion of C. J. Campbell from a vast amount of data and extensive consultation (*The Coming Oil Crisis,* 1997). A related difficulty will be the toxic residue that is even now saturating the biosystems of the planet, a residue that is devastating the forests of the planet and adversely affecting organic life everywhere. All this is in addition to the manner in which the toxic residue is affecting the drinking water of our urban centers.

North America has had a major role to play in the discovery and use of petroleum in the immense variety of its possibilities. After the first of the modern commercial oil wells was struck in the mid-nineteenth century in Pennsylvania, wells began to be sunk in other states, mostly along the Gulf of Mexico and across into California. We quickly learned that a great variety of products could be derived by refining the oil from its crude form as it came from the well. By the end of the century petroleum was being produced by countries in Eastern Europe as well as in Southeast Asia. The great oil fields of the Middle East were discovered in the late 1930s.

Throughout the twentieth century, the economic fortunes to be made from petroleum evoked a kind of frenzy for its discovery and use, especially after the automobile was invented and the roads were paved. Transportation has been the main use of the petroleum throughout the century, although petroleum derivatives soon began to be used for a vast number of petrochemicals as well as for manufacturing electricity.

Other sources of physical energy exist in the sunshine, the wind, the rivers, the tides, the forests. Yet in petroleum we have a manageable substance, easily transported, available on demand, that can provide the energy for the vast network of transportation facilities required by our modern economy. Petroleum can be stored for prolonged periods of time. It can fuel automobiles and aircraft. It can be transformed into fertilizers, pesticides, and herbicides for growing food, it can be spun into fibers for clothing, transformed into rubber for tires, made into asphalt for paving roads, shaped into numberless forms as plastics. It can provide the film and chemicals needed for photography. Thousands of products are now being made of petroleum and its derivatives. Thousands more are possible.

Considering all these uses, it is not surprising that the petroleum industry became the world's most profitable industry, supporting a vast number of other industrial establishments. Even agriculture, a foremost industry for its economic value, came to depend on petroleum for its fertilizer and for the machinery needed for sowing, cultivating, irrigating, harvesting, and marketing. Success in war depended heavily on the availability of petroleum and its products. Among the most lucrative of American corporations throughout the twentieth century have been the oil companies. The petroleum industry soon became the central feature of the world economy. Over a trillion dollars in business transactions concerned with petroleum were taking place annually by the 1980s. In 1998 when Exxon and Mobil Oil merged they formed the company with the largest assets of any existing industrial corporation.

The term *petrochemical* arises from the fact that petroleum is the basic resource from which the greater part of the chemicals now in use are produced. The invention of nylon from petroleum by DuPont in 1938 was the beginning of a new age in the weaving of fabrics and the fashioning of clothing. After 1909, plastics from petroleum came into modern industrial use. This was the beginning of the real magic of the petrochemical industry. Clear plastic could be shaped into something so delicate as eyeglasses. Plastics could readily be shaped into toys for children. Its use for packaging has become pervasive. In other contexts plastics could be given a hardness that would make it suitable for items formerly made of aluminum or steel. Petroleum was made into preservatives of many kinds as well as into varnishes and into dyes. In the form of glue it is used for making particleboards from wood chips.

Plastics have become so important to the medical profession that it would be difficult without plastics to carry out any of the remarkable procedures now performed by the medical profession. From the immense number of disposable gloves and gowns used by the profession each day, to the instruments it uses, to the tubes used in surgery, the medical profession is dependent on plastics. There are substitute materials to make plastics but none so easily available.

So many things came together in the last few decades of the nineteenth century to make the petroleum industry possible. The transition from craft technologies to scientific technologies was taking place, especially in the electrical and chemical industries. All of this required new developments in the sciences, especially in the fields of electrical and chemical engineering. The Rensselaer Polytechnical Institute, founded in 1837, was the first private institute in America to give a scientific engineering degree. Both physical and organic chemistry advanced in their understanding of the atomic structure of matter. The geological sciences advanced by the work of Sir Charles Lyell in 1833 and by Louis Agassiz in the 1850s.

Progress of the petroleum industry depended on all these developments in both the pure and in the applied sciences. Discovery of petroleum depended on geological studies. Bringing the crude oil to the surface required engineering skills. To refine crude oil into usable forms depended on the chemical and physical sciences and on the corresponding engineering skills.

At this time the automobile was still in the future. It was not long, however, before the modern world began to restructure itself under the influence of gasoline-powered automobiles and airplanes. Cities were transformed by the movement of people out into the once spacious suburbs. The inner cities began to decline as shopping centers and malls with their vast parking areas arose with access only by automobile. Developers made fortunes buying up land and then extending cities by building housing and shopping centers throughout the suburban areas.

After the main sources of petroleum obtained by present methods have been exhausted by the mid-twenty-first century, other sources more difficult to obtain and process, the oil sands and shale oil, could be exploited so that dwindling supplies could be available for another few decades. Yet exhaustion is inevitable. The proposal is being made that we can replace petroleum with hydrogen as an energy supply. This will certainly be at a significant economic cost. Even if we could supply ourselves with the energy needed, we would still be without easy replacements for the other uses to which petroleum is put. We can indeed make plastics, fabrics, and even fertilizers from chemicals available from coal or from other organic sources, but with much more difficulty and with greater pollution. It must always be remembered that in the dispositions of nature, petroleum will never again be made in any volume. The geobiological conditions by which petroleum came into being will never exist again.

We are urgently in need of a more sustainable setting for life here on the planet Earth, for even as the petroleum basis of life in the industrial world is declining, the developing nations are demanding a greater share in the resources of the world and all those products that constitute the distinctive way of life of industrial societies. Meanwhile in all our thinking about the future we must be conscious of the probable increase in the human population of the Earth from six billion to possibly eight to ten billion persons by the year 2050. Learning a way of life independent of petroleum for such a population may well be the most critical issue before the human community at the present time, because we cannot afford to wait until the crisis is upon us.

The difficulty of the petroleum period was that the well-being of the human was the final referent as regards reality and value. We could do so many things with petroleum that we began to have illusions concerning ourselves and our freedom to shape a world of our own that would challenge our dependence on nature. Everything on Earth, it seemed, got its value from its relation to the human. In reality the original design for Earth was for a planet that would be a triumph of diversity interacting with itself in a vast range of relationships. The well-being of each component part would be intimately related to the well-being of the other parts and to the well-being of the whole.

More than any other mode of being on the planet, humans were privileged to design themselves and their relationship with the other species through their cultural shaping. Other beings also would have a measure of self-shaping power, but to a somewhat lesser degree and in a somewhat different mode. Only humans seem to have the capacity to override the wide range of natural systems with their own self-devised systems. Evidence for this special ability is the amount of devastation that humans are bringing upon the planet at the present time.

This urgency of the need for a less toxic source for the benefits that chemistry now fashions out of a petroleum base is evident from the threat of a rising temperature of Earth, also from the threat of a thinning ozone layer above Earth, a layer that filters out some of the ultraviolet radiation from the sun that can cause significant damage to organisms if it reaches Earth in its unfiltered form. The order of magnitude of this planetary disturbance can be appreciated only if we consider the immense amount of time it has taken the universe to establish the physical conditions for the emergence of life in the variety of its present modes of expressions.

Here is where the modern industrial world reveals itself as failing in its larger purposes even while it seems to accomplish so much. It has failed to align its own functioning with the functioning of the planetary forces on which it depends. The intrusion of the chemical profession into the physics of the universe has enabled this profession to enter so profoundly into the hidden forces of the biological and the physical world that it can turn the most benign substances into the most deadly forms.

Fundamental to this attitude is the feeling that humans have the right and even the obligation to intrude upon the natural world as extensively as they are able. A person can only marvel that scientists generally seem never to have reflected on or explained to the community why the petroleum is buried in the Earth in the first place. Even the slightest reflection would reveal that nature has taken great care to bury the vast amounts of carbon in the coal and petroleum in the depths of the Earth and in the forests so that the chemistry of the atmosphere, the water, and the soil could be worked out with the proper precision. This needs to be thoroughly understood and respected lest anyone intrude into this delicate balance by extracting and using the petroleum or the coal or by cutting down and using the great forests of the planet without consideration of what will happen when these forces will no longer be able to fulfill their role in the integral functioning of the planet.

The petroleum interval is coming to its termination within the lifetime of persons living in the present. Yet there are still several decades when we can prepare for a future without petroleum. Of singular importance is the need to develop new forms of energy that are within the limits and restraints of nature's cycles. Some of these alternative sources, such as solar energy, radiant heat, and wind and water power have been identified by Amory and Hunter Lovins, John and Nancy Todd, and the Union of Concerned Scientists (Cole and Skerrett, 1995). Their research and writing are invaluable in pointing the way to a post-petroleum period. In the meantime, the best use for the available petroleum may be to use it carefully as we discover our way back to the Earth and learn how best to integrate a human way of life within the larger life community.

14

REINVENTING *the* HUMAN

WE MIGHT DESCRIBE THE CHALLENGE BEFORE US BY THE following sentence. The historical mission of our times is to reinvent the human—at the species level, with critical reflection, within the community of life-systems, in a time-developmental context, by means of story and shared dream experience.

I say *reinvent the human* because humans, more than any other living form, invent themselves. Other species receive their basic life instructions at the time of their birth. With this endowment they know how to obtain their food, how to defend themselves, how to obtain shelter, how to carry on their mating rituals, how to care for their young. Some species, especially the mammalian species, do need some teaching and guidance from an older generation. Young bears need to be taught how to fish. Many animal species need to be taught how to hunt. Yet this is minimal if compared to the extent of teaching and acculturation that humans need to arrive at maturity. That acculturation process is the purpose of the long childhood distinctive of humans. With their genetic endowment the other life-forms are, from an earlier age, much more fully equipped than humans to carry out their basic patterns of survival and propagation.

We need to reinvent the human *at the species level* because the issues we are concerned with seem to be beyond the competence of our present cultural traditions, either individually or collectively. What is needed is something beyond existing traditions to bring us back to the most fundamental aspect of the human: giving shape to ourselves. The human is at a cultural impasse. In our efforts to reduce the other-than-human components of the planet to subservience to our Western cultural expression, we have brought the entire set of life-systems of the planet, including the human, to an extremely dangerous situation. Radical new cultural forms are needed. These new cultural forms would place the human within the dynamics of the planet rather than place the planet within the dynamics of the human.

We must find our primary source of guidance in the inherent tendencies of our genetic coding. These tendencies are derived from the larger community of the Earth and eventually from the universe itself. In Jungian terms, these tendencies identify with those psychic energy constellations that take shape as the primary archetypal forms deep in the unconscious realms of the human. Such forms find expression in the symbols of the Heroic Journey, Death-Rebirth, the Sacred Center, the Great Mother, the Tree of Life. Although these symbolic forms are broadly the same in their general meaning, they each take on various modes of expression in the different religious and cultural traditions, modes that are analogously the same in their essential meaning.

The necessity of rethinking our situation at the species level is clear in every aspect of the human. As regards economics we need not simply a national or a global economy but local subsistence economies where the variety of human groups become acquainted with the other species in the local bioregion.

Our schools of business administration at the present time teach the skills whereby the greatest possible amount of natural resources is processed as quickly as possible, put through the consumer economy, and then passed on to the junk heap, where the remains are useless at best and at worst toxic to every living being. Now there is need for humans to develop reciprocal economic relationships with other life-forms providing a sustaining pattern of mutual support, as is the case with natural life-systems generally.

Especially as regards law, we need a jurisprudence that would provide for the legal rights of geological and biological as well as human components of the Earth community. A legal system exclusively for humans is not realistic. Habitat of all species, for instance, must be given legal status as sacred and inviolable.

Thirdly, I say *with critical reflection* because this reinventing of the human needs to be done with critical competence. Originally there was a certain instinctive, spontaneous process whereby the early cultural formations were established. Now we need all our scientific insight and technological skills. We must, however, see that our sciences and technologies are coherent with the technologies of the natural world. Our knowledge needs to be in harmony with the natural world rather than a domination of the natural world. We need the art of intimate communion with, as well as technical knowledge of, the various components of the natural world.

We insist on the need for critical reflection as we enter the ecological age in order to avoid a romantic attraction to the natural world that would not meet the urgencies of what we are about. The natural world is violent and dangerous as well as serene and benign. Our intimacies with the natural world must not conceal the fact that we are engaged in a constant struggle with natural forces. Life has its bitter and burdensome aspects at all levels. Yet its total effect is to strengthen the inner substance of the living world and provide the never-ending excitement of a grand adventure.

Fourth, we need to reinvent the human *within the community of life systems.* This is the central phrase, the primary condition for reinventing the human. Because the Earth is not adequately understood either by our spiritual or by our scientific traditions, the human has become an addendum or an intrusion. We have found this situation to our liking since it enables us to avoid the problem of integral presence to the Earth. This attitude prevents us from considering the Earth as a single community with ethical relations determined primarily by the well-being of the total Earth community.

While the Earth is a single community, it is not a global sameness. Earth is highly differentiated into arctic as well as tropical regions, into mountains, valleys, plains, and coastlands. These geographical regions are also bioregions. Such areas can be described as identifiable geographical regions of interacting life-systems that are relatively self-sustaining in the ever-renewing processes of nature. As the functional units of the planet, these bioregions can be described as self-propagating, self-nourishing, self-educating, self-governing, self-healing, and self-fulfilling communities. Human population levels, our economic activities, our educational processes, our governance, our healing, our fulfillment must be envisaged as integral with this community process.

There are great difficulties in identifying just how to establish a viable context for a flourishing and sustainable human mode of being. Of one thing we can be sure, however: our own future is inseparable from the future of the larger community that brought us into being and sustains us in every expression of our human quality of life, in our aesthetic and emotional sensitivities, our intellectual perceptions, our sense of the divine, as well as in our physical nourishment and our bodily healing.

Fifth, reinventing the human must take place *in a time-developmental context.* This constitutes what might be called the cosmological-historical dimension of the program I am outlining here. Our sense of who we are and what our role is must begin where the universe begins. Not only does our physical shaping and our spiritual perception begin with the origin of the universe, so too does the formation of every being in the universe. This human formation is governed by three basic principles: differentiation, subjectivity, and communion.

Our present course is a violation of each of these three principles in their most primordial expression. Whereas the basic direction of the evolutionary process is toward constant differentiation within a functional order of things, our modern world is directed toward monocultures. This is the inherent direction of the entire industrial age. Industry requires a standardization, an invariant process of multiplication with no enrichment of meaning. In an acceptable cultural context, we would recognize that the unique properties of each reality determine its value both for the individual and for the community. These are fulfilled in each other. Violation of the individual is an assault on the community.

As a second imperative derived from the cosmological process, we find that each individual is not only different from every other being in the universe but also has its own inner articulation, its unique spontaneities. Each being in its subjective depths carries that numinous mystery whence the universe emerges into being. This we might identify as the sacred depth of the individual.

The third cosmological imperative indicates that the entire universe is bonded together in such a way that the presence of each individual is felt throughout its entire spatial and temporal range. This capacity for bonding the components of the universe with one another enables the vast variety of beings to come into existence in that gorgeous profusion and yet comprehensive unity we observe about us.

From this we can appreciate the directing and energizing role played by the *story* of the universe. This story that we know through empirical observation of the world is our most valuable resource in establishing a viable mode of being for the human species as well as for all those stupendous life-systems whereby the Earth achieves its grandeur, its fertility, and its capacity for endless self-renewal.

This story, as told in its galactic expansion, its Earth formation, its life emergence, and its self-reflexive consciousness, fulfills in our times the role of the mythic accounts of the universe that existed in earlier times, when human awareness was dominated by a spatial mode of consciousness. We have moved from cosmos to cosmogenesis, from the mandala journey toward the center of an abiding world to the irreversible journey of the universe itself, as the primary sacred journey. This journey of the universe is the journey of each individual being in the universe. So this story of the great journey is an exciting story that gives us our macrophase identity with the larger dimensions of meaning that we need. To identify the microphase of our being with the macrophase mode of our being is the quintessence of human fulfillment.

The present imperative of the human is that this journey continue on into the future in the integrity of the unfolding life-systems of the Earth, which presently are threatened in their survival. Our great failure is the termination of the journey for so many of the most brilliant species of the life community. The horrendous fact is that we are, as the Cambridge University biologist Norman Myers has indicated, in an extinction spasm that is likely to produce "the greatest single setback to life's abundance and diversity since the first flickerings of life almost four billion years ago" (quoted in *Biodiversity*, p. 34). The labor and care expended over some billions of years and untold billions of experiments to bring forth such a gorgeous Earth is all being negated within less than a century for what we consider "progress" toward a better life in a better world.

The final aspect of our statement concerning the ethical imperative of our times is *by means of the shared dream experience*. The creative process, whether in the human or the cosmological order, is too mysterious for easy explanation. Yet we all have experience of creative activity. Since human processes involve much trial and error with only occasional success at any high level of distinction, we may well believe that the cosmological process has also passed through a vast period of experimentation in order to achieve the ordered processes of our present universe. In both instances something is perceived in a dim and uncertain manner, something radiant with meaning that draws us on to a further clarification of our understanding and our activity. This process can be described in many ways, as a groping or as a feeling or imaginative process. The most appropriate way of describing this process seems to be that of dream realization. The universe seems to be the fulfillment of something so highly imaginative and so overwhelming that it must have been dreamed into existence.

But if the dream is creative we must also recognize that few things are so destructive as a dream or entrancement that has lost the integrity of its meaning and entered an exaggerated and destructive manifestation. This has happened often enough with political ideologies and religious visionaries. Yet there is no dream or entrancement in the history of the Earth that has wrought the destruction that is taking place in the entrancement with industrial civilization. Such entrancement must be considered as a profound cultural disorientation. It can be dealt with only by a corresponding deep cultural therapy.

Such is our present situation. We are involved not simply with an ethical issue but with a disturbance sanctioned by the very structures of the culture itself in its present phase. The governing dream of the twentieth century appears as a kind of ultimate manifestation of that deep inner rage of Western society against its earthly condition as a vital member of the life community. As with the goose that laid the golden egg, so the Earth is assaulted in a vain effort to possess not simply the magnificent fruits of the Earth but the power itself whereby these splendors have emerged.

At such a moment a new revelatory experience is needed, an experience wherein human consciousness awakens to the grandeur and sacred quality of the Earth process. This awakening is our human participation in the dream of the Earth, the dream that is carried in its integrity not in any of Earth's cultural expressions but in the depths of our genetic coding. Therein the Earth functions at a depth beyond our capacity for active thought. We can only be sensitized to what is being revealed to us. We probably have not had such participation in the dream of the Earth since earlier shamanic times, but therein lies our hope for the future for ourselves and for the entire Earth community.

15

The DYNAMICS *of the* FUTURE

AS WE ENTER THE TWENTY-FIRST CENTURY WE OBSERVE A widespread awakening to the wonder of the Earth. This we can observe in the writings of naturalists and the environmental organizations dedicated to preserving the integrity of the planet. There are also those in the scientific world who give expression to the wonder of things, such as Peter Raven, Norman Myers, Lynn Margulis, Eric Chaisson, Ursula Goodenough, Brian Swimme, and others who are revealing to us the larger pattern as well as the intricate details of the visible world about us.

The human venture depends absolutely on this quality of awe and reverence and joy in the Earth and all that lives and grows upon the Earth. As soon as we isolate ourselves from these currents of life and from the profound mood that these engender within us, then our basic life-satisfactions are diminished. None of our machine-made products, none of our computer-based achievements can evoke that total commitment to life from the subconscious regions of our being that is needed to sustain the Earth and carry both ourselves and the integral Earth community into the hazardous future.

How we feel about ourselves and about the Earth process are questions of utmost urgency, especially when we are presented with the notion of Earth as a collection of commodities to be bought and sold. The very meaning of the Earth is involved in this question, as are the human energies needed to assist in shaping a desirable future. In our quest for understanding, we might begin with the observation that the Earth is the manifestation of a vast amount of energy caught up in a diversity of designs for which there is no accounting in terms of human understanding or imagination. In a sequence of mutations great stores of energy were deposited within the Earth, not only as fossil fuels but also as life forces within the very structure of matter.

Our present peril is not the first that the Earth and living things upon it have endured. The Earth found its way into being amid an amazing sequence of both creative and destructive experiences. A long sequence of cataclysmic events has shaped the continents and the various forms of life that have themselves engaged in a continuing struggle for survival. But the present danger to the planet is the first conscious intrusion on this scale into the natural rhythms of the Earth process. This is something radically different from the seismic convulsions, the glacial advances, the earlier emergence and disappearance of species. It is the exploitation of energies in a definitive form. It is a turn from the storing of energies to the burning off of energies in a manner and in a volume such that they cannot be replaced in any conceivable period of human historical existence. Because of our need to fuel the industrial world, we have created a technosphere incompatible with the biosphere.

In the late Paleolithic and the Neolithic periods, also in the earlier stages of the classical civilizations, we lived in an ocean of energy. Physical and psychic forms of energy were intimately related. We found the meaning of existence as we responded to the energies about us.

These we perceived ultimately as spirit forces. During this period the resources of Earth were little diminished. Although humans put stress upon the planet, they lived necessarily in a certain harmony with the Earth process. The natural rhythms were only slightly disturbed. From the unconscious depths of the human psyche came the great visions. A feeling of identity with the Earth was at its height. Earth was the Great Mother, heaven a comprehensive Providence. The ultimate mystery of things was celebrated in ritual worship. This was "the Age of the Gods," as Giambattista Vico termed it in the early eighteenth century. Zeus was a presiding presence. Poseidon ruled the seas. Athena inspired wisdom, Dionysus promoted wine-drinking, music, and festivals, Venus excited love, Mars gave courage in war, Demeter brought abundant harvest, Diana and the nymphs gave spirit presence to the woodlands. The constellations in the night sky gave meaning to the night. The universe was the expression of such personal and spirit powers.

Human psychic energies as well as Earth's physical energies were stored up in such abundance that all later ages have found in this period their primordial cultural support. Despite all the changes that have taken place in later times, this age still holds many of our normative values. We feel the support of these mythic values, although the historical realism of Western civilization and our later scientific empiricism have weakened the conscious presence to the natural world that once guided and supported the psychic dimension of our lives.

In more recent times we have replaced these earlier mythic structures first with the myth of continuing and inevitable progress, then with the myth of evolution. While this sense of necessary and inevitable progress originated in the millennial promises of the Book of Revelations it was developed and expanded in the modern period with the new sciences that came into being. Originally it was the progress of the human mind that dominated the early modern period. In the *Novum Organum* (1620), *The New Method of Learning*, Francis Bacon proposed the empirical mode of reasoning to replace the deductive method of the medieval period. What Bacon proposed Galileo Galilei carried out in his experiments with motion and in his observation of the heavens. These were accompanied with development of the new technologies of the telescope and microscope. In *The New Science of the Nature of the Nations* (1725), Giambattista Vico proposed a new way of understanding the historical unfolding of the social and cultural orders of life. Bernard Fontinelle and George Buffon, in this same period, fostered a recognition that the Earth was much older than the five thousand years indicated by biblical reckoning. In their estimate it must be at least eighty thousand years old. Denis Diderot and Jean d'Alembert edited the great French *Encyclopédie* that appeared between 1752–1772 in twenty-eight volumes, the first of the comprehensive modern encyclopedias. This encyclopedia became the main referent for exposition of the new rationalism and the doctrine of progress.

This myth of progress supplanted the earlier myths of personal presences manifested throughout the natural world. At this same time we lost the world of meaning in an evolutionary world governed by chance without direction or higher significance, a world of emergent process that would eventually come to be spoken of as the work of a "blind watchmaker," as in Richard Dawkins's book *The Blind Watchmaker.* Yet a different interpretation of the data of evolution is available. We need merely understand that the evolutionary process is neither random nor determined but creative. It follows the general pattern of all creativity. While there is no way of fully understanding the origin moment of the universe we can appreciate the direction of evolution in its larger arc of development as moving from lesser to great complexity in structure and from lesser to greater modes of consciousness. We can also understand the governing principles of evolution in terms of its three movements toward differentiation, inner spontaneity, and comprehensive bonding.

With this understanding it would be difficult to overemphasize the magnificence of this evolutionary doctrine. It provides a grandeur in our view of the universe and our human role in it that is overwhelming. Indeed, in its human expression the universe is able to reflect on itself and enjoy its grandeur in a special mode of conscious self-awareness. The evolutionary vision provides the most profound mystique of the universe.

Our main source of psychic energy in the future will depend on our ability to understand this symbol of evolution in an acceptable context of interpretation. Only in the context of an emergent universe will the human project come to an integral understanding of itself. We must, however, come to experience the universe in its psychic as well as in its physical aspect. We need to experience the sequence of evolutionary transformations as moments of grace, and also as celebration moments in our new experience of the sacred.

As physical resources become less available, psychic energy must support the human project in a special manner. This situation brings us to a new reliance on powers within the universe and also to experience of the deeper self. The universe must be experienced as the Great Self. Each is fulfilled in the other: the Great Self is fulfilled in the individual self, the individual self is fulfilled in the Great Self. Alienation is overcome as soon as we experience this surge of energy from the source that has brought the universe through the centuries. New fields of energy become available to support the human venture. These new energies find expression and support in celebration. For in the end the universe can only be explained in terms of celebration. It is all an exuberant expression of existence itself.

This sense of celebration brings us back to the earlier pages of this book, where I mentioned the manner in which indigenous peoples and the peoples who shaped the earlier civilizations sought to coordinate every human activity with the seasonal renewal moments of the natural world. Every phase of life found validation in this larger context. But while in former times we celebrated the moments of seasonal renewal, now we must also celebrate the sequential transformation moments in an emergent universe. This story of the emergent universe is now our dominant sacred story.

The Sundance of the Lakota Sioux tribe in North America provides an instance of celebration and renewal founded in the seasonal renewal experience. This dance is one of the most dramatic forms of cosmological integration of the human with the universe. The center pole is seen as a place for the intersection of the numinous, the cosmic, and the human worlds. Through this ceremony the dancers attain a heroic consciousness combined with a sense of having rendered a debt of gratitude for the gift of life granted the individual and the tribe. A certain psychic energy is developed both within the individual and within the tribe, an interior energy that enables the community to accept the difficulties of life with an endurance and an equanimity that has ever amazed the Euro-Americans. While we need to continue such seasonal-based liturgies like those of indigenous peoples and of the classical traditions, we need to create new liturgies based on the transformation episodes of the evolutionary sequence in the physical, geological, and biological stages of development. In both cases the inner psychic resources of the human venture are restored and increased.

There is a significant difference between physical energy and psychic energy. Physical energy is diminished by use. Once used, we are left with lifeless matter and waste products that are often dangerous to the life-process. In contrast, psychic energies are increased by use and by the numbers of those who participate in their activity. Material things are diminished as more people share them while nonmaterial realities are enhanced as more people share them. If a piece of food is shared by four people each gets much more than the same food shared by ten people. Understanding, joy, spiritual insight, music, and the arts, however, are augmented as the audience is increased or as they are communicated from one person to another. A completely individual emotion is hardly of the human order. It needs to be shared. In the sharing, the resonance is increased and a greater range of human experience is established.

Along with a new appreciation of the psychic dimension of life, we now see a strengthening of the ancient symbols of the human community, which lead to a deeper experience of the self. Through the archetypal symbols of the unconscious, the symbols of the Great

Journey, Death-Rebirth, the Cosmic Tree, and the Sacred Center we establish vital contact with those underlying energies that guide and sustain the human soul. These energies find expression especially in our dreams, for it is those larger dreams of a civilization that guide and drive the actions of a people. The symbols of the past now have a range of meaning that they never had previously, for now the well-being and the survival of the planet, and all the other modes of being on the planet, are involved in our human actions.

Previously our actions were of critical concern only for some phase of the human venture. Through these symbols interior energies that sustain the human venture and the entire civilizational process are renewed. Ultimately these symbols reflect the powers of the Earth both in their ever-renewing seasonal significance and in their irreversible evolutionary transformations. Both of these make it possible for human understanding and intuition to be fully responsive to the cosmological processes that brought us into being and sustain us in existence.

Each of the symbols we have mentioned has a new richness of interpretation. The journey symbol is no longer simply the journey from the circumference to the center within the context of the mandala where the divine, the human, and the cosmos become present to each other. The journey must now be understood also as the great journey that the universe has made from its primordial flaring forth until the present. This journey is carried out through a new mode of presence of these three to one another.

So too the Death-Rebirth symbol is not simply the death-rebirth of our traditional initiation symbolism whereby we move from a profane to a sacred form of life. The Death-Rebirth symbol must now be understood in relation to the moments when the universe undergoes some significant transformation whereby one geological or biological period moves into a greater manifestation of the numinous mystery whence all phenomena come into being. A similar twofold significance must be understood of the other archetypal symbols that govern human activities from deep within the unconscious depths of the human soul. All former symbols must acquire a new depth of meaning without diminishing their prior meaning.

The other force that needs to be mentioned, wherein we might place our hopes for the future, is the force of will. Although this subject of will was the principal concern of Arthur Schopenhauer (1788–1860) in the nineteenth century, it has not been adequately dealt with either speculatively or practically in recent times. Yet seldom have such demands been made on the will as are being made at the present time. A concern for will does appear in the work of Pierre Teilhard de Chardin, particularly in *Human Energy.* He saw quite clearly that we must consciously will the further stages of the evolutionary process. Almost immediately this responsibility became too much for us to carry. We live now in a moment of indecision, carrying the world in our hands, afraid of tripping over our own feet and letting it fall to its destruction.

But while this peril is a cause for concern, it is also a cause for advancing consciousness. Responsible people no longer think of the world simply as a collection of natural resources. We have begun to realize that the Earth is an awesome mystery, ultimately as fragile as we ourselves are fragile. But our responsibility to the Earth is not simply to preserve it, it is to be present to the Earth in its next sequence of transformations. While we were unknowingly carried through the evolutionary process in former centuries, the time has come when we must in some sense guide and energize the process ourselves.

To succeed in this task of shaping the future, the will of the more comprehensive self must be functioning. The individual will can function in this capacity only through an acknowledged union with the deeper structures of reality. Even beyond union with the human community must be union with the Earth, with the universe itself in the full wonder of its being. Only the Earth can adequately will the Earth. If we will the future effectively it will be because the guidance and the powers of the Earth have been communicated to us, not because we have determined the future of the Earth simply with some rational faculty.

Central to this process is our contact with the sacred and the vast range of Earth's psychic dynamism. While our sense of the sacred can never be recovered precisely as it existed in former centuries, it can be recovered in the mystique of the Earth, in the epic of evolution. Spiritual disciplines are once again being renewed throughout the world. For some the ultimate mystery of things is experienced in the depth of the inner self, for others in the human community, for still others in the Earth process itself. Yet in each instance the full sense of communion seems to be present. A way is opening for each person to receive the total spiritual heritage of the human community as well as the total spiritual heritage of the universe. Within this context the religious antagonisms of the past can be overcome, the particular traditions can be vitalized, and the feeling of presence to a sacred universe can appear once more to dynamize and sustain human affairs.

We must feel that we are supported by that same power that brought the Earth into being, that power that spun the galaxies into space, that lit the sun and brought the moon into its orbit. That is the power by which living forms grew up out of the Earth and came to a special mode of reflexive consciousness in the human. This is the force that brought us through more than a million years of wandering as hunters and gatherers; this is that same vitality that led to the establishment of our cities and inspired the thinkers, artists, and poets of the ages. Those same forces are still present; indeed, we might feel their impact at this time and understand that we are not isolated in the chill of space with the burden of the future upon us and without the aid of any other power.

We are a pervasive presence. By definition we are that reality in whom the entire Earth comes to a special mode of reflexive consciousness. We are ourselves a mystical quality of the Earth, a unifying principle, an integration of the various polarities of the material and the spiritual, the physical and the psychic, the natural and the artistic, the intuitive and the scientific. We are the unity in which all these inhere and achieve a special mode of functioning. In this way the human acts as a pervading logos. If the human is microcosmos, the cosmos is macroanthropos. We are each the cosmic person, the *Mahapurusha,* the Great Person of Hindu India, expressed in the universe itself.

This being so, there is need to be sensitive to the Earth, for the destiny of the Earth identifies with our own destiny, exploitation of the Earth is exploitation of the human, elimination of the aesthetic splendors of the Earth is diminishment of existence. We do not serve the human by blasting the mountains apart for mineral resources, for in losing the wonder and awesome qualities of the mountains we destroy an urgent dimension of our own reality.

Ancient rituals through which we communicated with the Earth and fostered its productivity may no longer seem fully effective. Yet they do express a profound respect for the mystery of the Earth. It would be philosophically unrealistic, historically inaccurate, and scientifically unwarranted to say that the human and the Earth no longer have an intimate and reciprocal emotional relationship.

We are not lacking in the dynamic forces needed to create the future. We live immersed in a sea of energy beyond all comprehension. But this energy, in an ultimate sense, is ours not by domination but by invocation.

16

The FOURFOLD WISDOM

IN THESE OPENING YEARS OF THE TWENTY-FIRST CENTURY, AS the human community experiences a rather difficult situation in its relation with the natural world, we might reflect that a fourfold wisdom is available to guide us into the future: the wisdom of indigenous peoples, the wisdom of women, the wisdom of the classical traditions, and the wisdom of science. We need to consider these wisdom traditions in terms of their distinctive functioning, in the historical periods of their florescence, and in their common support for the emerging age when humans will be a mutually enhancing presence on the Earth.

Indigenous wisdom, which extends far back into the Paleolithic Period, survives even into the present among the 200 million indigenous peoples. The wisdom of women, which flourished throughout the Neolithic Period, is now experiencing a reassertion of itself in a new form. The wisdom of the urban classical literate traditions had its beginning some five thousand years ago and was the most powerful force in human cultural formation until it was challenged by the scientific tradition of more recent times. Science as a wisdom tradition is only in its beginning phase, even though scientific knowledge has advanced with amazing success ever since the sixteenth century.

INDIGENOUS WISDOM is distinguished by its intimacy with and participation in the functioning of the natural world. The dawn and sunset are moments when the numinous source of all existence is experienced with special sensitivity. In springtime the flowering world sets forth its blossoms. The birds appear in the brilliance of their coloring, in the ease and skill of their flight, and in the beauty of their song. Then there are the fearsome moments when thunder rolls across the heavens and the lightning shatters both the sky and the Earth. In the Northern Hemisphere when autumn comes the fruits appear, the birds depart, the leaves fall, darkness settles over the land. In the various tropical regions the rains come more fully in their seasonal sequence. Living things come into being, flourish, then fade from the scene. This ever-renewing sequence of sunrise and sunset, of seasonal succession, constitutes a pattern of life, a great liturgy, a celebration of existence.

In this context early humans discovered their food and sheltered themselves from the elements. Above all they developed an interpretation of life and pain and suffering and death, along with a feeling of security and joy in existence. A native wisdom was passed down through the generations, a wisdom carried in the lives of the people, in their thoughts and speech; in their customs, songs, and rituals; in their arts, their poetry; and in their stories not in any written form. In a special manner this wisdom is carried by the sacred personalities: the elders—both men and women, the chiefs, the shamans.

One of the primary accomplishments of humans in the early millennia of their existence was the occupation of the various continents of the world. With amazing genius they developed spoken language. During this late Paleolithic and throughout the Neolithic periods thousands of languages were invented. In Africa and in South America alone several thousand languages have existed into recent times. The primary technologies were invented. Indigenous peoples learned the art of making fire, fashioned elementary weapons, identified plants and animals, especially those plants associated with food and with healing. They shaped stone instruments needed in their hunting culture. Such creative relationships with the natural world resulted from an intimacy with that world.

The arts, music, and literature were also developed in this early period, especially in the late Paleolithic and into the Neolithic. While we are acquainted with these mainly in their later more highly articulated form, as we find them in their early Neolithic phases, we can recognize that these later forms are themselves the consequence of a long period of development. In many of their accomplishments, such as the wood carvings of African peoples and in the bark paintings of the Aborigines of Australia, we experience a powerful spirit presence. We also experience the vast range of thought, history, and spiritual insight contained in these forms.

Such sensitivity to the powers of the universe resulted from the experience of benign and fearsome forces surrounding the human community. In America the pervasive spirit presence was experienced as essentially benign. This sense of a benign universe found expression in the Navaho expression to "go in beauty." Beauty is a way of expressing that numinous presence that is pervasive throughout the universe. It expresses one of the ways in which this spirit presence is given identity. Awareness of this spirit presence is evident throughout the ceremonial lives of tribal peoples throughout the world.

These powers in back of all natural phenomena are seen as personal in nature. They are named and addressed as modes of conscious presence presiding over the functioning of the universe both in its comprehensive extent and in its individual intimacy. We see this intimate way of relating to the world in the Bushmen of Africa. As indicated by Laurens van der Post in *A Far-Off Place*, the teacher of a young boy insists on reverence for the forest: "One is never alone in the forest. One is never unobserved" (p. 79).

The great advantage in the modern European contact with the indigenous peoples of the world is the perspective that it has provided peoples of Western European civilization with an occasion to reflect on the inherent consequences of the civilizational process itself. For the first time, in the beginning of the colonial period, Western civilization could be seen as being weakened, both physically and morally, precisely through the civilization process itself. The primordial vigor, the fundamental heroic virtues, tend to survive only in diminished form. So too we tend to see other peoples either in a situation of savage primitivity or of a romantic primordialness.

In more recent times we have come to know the way of life of the indigenous peoples of the world in more extensive detail. Yet we somehow remain estranged from the deeper realms of their consciousness. As the years pass, however, we have begun to recognize both how little we really understand these peoples and how much we need the wisdom of their traditions. We have been instructed by an ever-increasing number of native persons fully capable of speaking the wisdom of their heritage as this comes down from earlier times into the present.

There is, correspondingly, a greater willingness and a greater ability to understand which is manifested by peoples of other traditions now living on this continent. We might hope that a capacity for communication is also developing. Awareness of past dispossession of the Indian lands is increasing. Yet, if devastation of their former way of life by hunting and fishing and gathering and destruction of their more elaborate cultural achievements has taken place in what is sometimes an irreversible manner, the Indian peoples of this continent do possess, it seems, an indestructible psychic formation that will remain into the indefinite future. They have held on to dimensions of their ancient wisdom traditions of which European Americans have had little or no knowledge.

As the years pass it becomes ever more clear that dialogue with native peoples here and throughout the world is urgently needed to provide the human community with models of a more integral human presence to the Earth.

THE WISDOM of women is to join the knowing of the body to that of mind, to join soul to spirit, intuition to reasoning, feeling consciousness to intellectual analysis, intimacy to detachment, subjective presence to objective distance. When these functions become separated in carrying out the human project then the way into the future is to bring them together.

The human project belongs to both women and men. It cannot be carried out effectively unless both are present throughout the full range of human activities: with each other and with the family, in government, in the economic enterprise, in university education, in the religious establishment, in the artistic and literary worlds. Wherever the human enterprise is taking place women belong there as well as men. Each brings distinctive abilities to the single project.

Because men in Western civilization have isolated women in the home and in a narrow band of service activities, and have appropriated for themselves both the reality and value of the adult human outside the home, the human project in its Western manifestation has become a patriarchal establishment in quest for unlimited dominance, a dominance unsettled in itself and a disturbance to the larger human community. In a corresponding way, because men have appropriated the reality and value of the Earth for their own purposes, the Earth is becoming dysfunctional. Again it is a quest for dominance.

The only acceptable attitude of any mode of being is to recognize that existence is a mutual dependence of a diversity of components. The human is a single enterprise that brings together women and men, elders and children, the farmer and the merchant, the foreigner and the native. So too the Earth is a single enterprise, composed of land and sea, rain and wind, plants and animals and humans, and the whole magnificent universe. Nothing is itself without everything else. This centering of men on themselves, to the detriment of women, the home, and the family as well as the Earth and everything on the Earth, is identified as *androcentrism.*

Among the greatest of terrors that women have encountered as a consequence of androcentrism is when they are considered the result of a genetic deficiency, are thought of as intellectually deficient, are seen as inherently seductive in their moral influence, are persecuted as malevolent spirits. In some societies women are required to undergo physical mutilation, are sold or forced into prostitution, or are traded off in arranged marriages. An endless list could be compiled of abuses and suppressions imposed upon women in the past and that continue even in the present, especially in the form of sexual exploitation. Women were thought to exist for the purposes of men.

By asserting their place in every aspect of the social and cultural life of the human community, women are bringing their own resolution to this attitude of men. While women are thus fulfilling a duty they owe to themselves, they are also revealing to men the reality of the patriarchal dominance that men have been imposing on the human community. Women are also revealing Western civilization to itself. Without this newly assertive consciousness of women, Western civilization might have continued indefinitely on its destructive path without ever coming to a realization of just what has been happening in the exclusion of women from full participation in the human project.

This revelation of men to themselves and of Western civilization to itself might be considered the first most dramatic manifestation of the wisdom of women. The transformation of men and of Western civilization is a primary condition for every other change that is needed in shaping a future worthy of either men or women. Androcentrism and patriarchy bring down in ruins the finest aspirations of the religious and humanist traditions of Western civilization—and also, it seems, most other civilizations of the Eurasian world that have dominated the first several thousand years of the modern human project. It might even be indicated that the foundations of our cultural traditions have from the beginning tended in this direction. A truly realistic insight into the situation might reveal that the distortions we mention might be less a deviation than the fulfillment of certain aspects of the Western tradition. This new interpretation of Western history, as dominated by its patriarchal establishments, can be considered the most profound contribution to historical understanding made in recent centuries.

The first concern just now is for men to accept the transformation in Western civilization indicated by abandoning their androcentrism and their patriarchal dominance. This transformation is a historical task forced upon the society by women. Men can best assist by welcoming the transformation being effected and by recognizing their responsibility for the burden that women have endured through these past centuries, and by understanding what women are communicating to them.

One of the first steps for women now is assertion of their role within the larger realm of human affairs. Assertion of their personal dignity and their personal rights in the socioeconomic order is an elementary first step that needs to be taken. Distinguished women in every sphere of life and in all the professions can be identified throughout history, but especially in the nineteenth and twentieth centuries. With Madame Marie Curie (1867–1934), women attained distinction in the scientific realm. She was followed by many other women who entered the scientific as well as the various professional occupations: Maria Montessori in pedagogy (1870–1952), Rachel Carson (1907–1964) in biology, Barbara McClintock (1902–1992) in understanding the genetic process.

While there are innumerable stories of women in their public careers and in their professional achievements, there exists also, throughout the twentieth century, a long list of women writers and activists more directly concerned with fostering the personal, the social, and the professional well-being of women. The nature and the extent of change that women will initiate in the coming years throughout the social and cultural context of contemporary life is beyond imagining. Already, though, Mary Joy Breton has listed the contributions that women are making to the environmental renewal of the Earth in her book, *Women Pioneers for the Environment.* Until recently men seem to have attained more prominence in nature writing, women seem to be more prominant in active doing, especially in the immediate work of preserving a viable planet for future generations.

Out of their historical experience an immense store of wisdom is available to women for influencing the course of the future in its every aspect, from its social and cultural through its religious institutions, its educational establishments, and its economic functioning. Yet there is, it seems, a deeper background to the identity of woman, a background that reaches far back in time to the Neolithic Period, a background that is of special significance in relation to the task of moving the human project into a period when humans would be present to the Earth in a mutually enhancing manner.

The earliest and most profound human experience of woman in these former centuries of human development is found in identifying maternal nurturance as the primordial creating, sustaining, and fulfilling power of the universe. Mutual nurturance is presented as the primary bonding of each component of the universe with the other components. This experience of the universe as originating in and sustained by a primordial originating and nurturing principle imminent in the universe itself finds expression in the figure of the Goddess in the late Paleolithic Period and in the Neolithic Period in the Near East.

This Goddess figure presided over this period as a world of meaning, of security, of creativity in all its forms. This was not a matriarchy, nor was it a social program. It was a comprehensive cosmology of a creative and nurturing principle independent of any associated male figure. This is difficult for us to envisage, a world with a comprehensive cosmology of woman with a derivative sociology. It is a tribute to contemporary women that they are developing both a sociology and a cosmology of woman.

The records of this early period come down to us in archaeological remains with an immense number of figurines and symbolic impressions in clay. These have finally been excavated and interpreted for us by Marija Gimbutas. Although we now know this period more thoroughly than we have ever known it previously, the historical studies have, until recently, been overwhelmed by both the archaeological and the historical studies of the next period, the period dominated by the male warrior deities that presided over the religious life of the period from 3000 BCE. Only now have these earlier sources from the Neolithic Period been fully studied by a woman. Gimbutas has found and interpreted what, it seems, no one else has been able to fully appreciate.

From these archaeological studies of the Neolithic Period, also from historical references of the succeeding age, we can conclude that this period of the Goddess was a relatively peaceful time, of intimate human presence to the Earth and to the entire natural world. The first permanent villages were established. The first domestication of plants and animals took place in these settled communities. We are just beginning to appreciate how creative a time it was when the Goddess culture had a pervasive presence throughout human affairs. So far scholars have been mainly concerned with this period for what it meant in terms of human capacity to shape village communities and to exploit the land by growing crops.

The male deities did not rise out of this context. These frequently warrior deities arose from without and became dominant throughout the Eurasian landmass as Zeus, Yahweh, Indra, and Thor. All of this was the beginning of the urban, literate, patriarchal civilizations, with the subordination of feminine deities in the Greek world and the denunciation of feminine deities in the biblical world. Once this dominant position of men as divine rulers or as rulers with divine sanction was established, a fixation developed that could not be successfully altered until the religious traditions of Western civilization and their associated deity came to be challenged in their basic meaning.

THE WISDOM of the classical traditions is based on revelatory experiences of a spiritual realm both transcendent to and imminent in the visible world about us and in the capacity of humans to participate in that world to achieve the fullness of their own mode of being. The Hindu tradition of India is founded on the revelation of the unity of the deepest self of the universe, the *Atman,* with the inner self of the human, a revelation received by the ancient *rishis* in the Upanishadic meditations. Expressed in the phrase "Thou art That" the individual self finds its identity in the Great Self. The Buddhist tradition is founded on the enlightenment experience of Gautama Buddha, for whom the universe was revealed as transient, sorrowful, and unreal. The immediate conclusion is expressed in the teaching of universal compassion. In their final fulfillment every being participates in the primordial reality of the Buddha. The Chinese experience moves less toward the transcendent world than to the spontaneities and the inner rhythms of the cosmos. The supreme experience is that of unity with the "One Body" of all the components of the universe, "the ten thousand things."

The classical wisdom of the Western world is a wisdom focused on the existence of a monotheistic personal male deity, creator of the universe clearly distinct from himself, a deity communicating his directions for the human community to a small pastoral tribe in the Palestinian region of the eastern shores of the Mediterranean Sea. The personalities in communication with this deity were considered as prophetic personalities, persons who speak for the deity.

The main characteristic of this primordial wisdom of the Western world is its direct communication from this supreme personal deity, who later appeared in human form as teacher and as savior. The revelation communicated for all the peoples of the world includes a pattern of historical achievement to be accomplished by a chosen group of believers. The historical dynamism of this tradition has driven the course of the Western world down through the centuries with the conviction that it is leading the entire human community to fulfillment in a divine kingdom, a kingdom with a millennial fulfillment on Earth in historical time and a posthistorical fulfillment in an eternal transcendental mode of being.

Another major component of Western tradition is the Greek humanist tradition as expressed in their literary and artistic accomplishments; also as expressed in the tragedies of Aeschylus and Sophocles, in the teachings of the philosophers, especially of Plato and Aristotle, and in the architecture and sculptures of the Parthenon. When the biblical tradition and the Greek humanist traditions came together, the full power of the thought and spiritual tradition of the Western world began to take shape.

In the early fourth century the Roman Empire became officially Christian. Yet it required a long period of assimilation and long centuries of internal and external strife before components of Western civilization could come together in a coherent cultural expression that eventually gave rise to medieval Europe.

When these three traditions—the biblical, the Greek humanist, and the Roman imperial—came together and the conversion of the barbarian tribes began to occur, the tradition that would become the dominant power in future world history was assembled in all its basic components. It might even be said that a force was assembled that one day would pass beyond the merely human world and seek to impose its technological dominion upon the natural world, with the consequence of disrupting the geological and biological systems of the Earth. Even though the religious tradition would be rejected in this later scientific age, the deep mind formation of the scientist, as well as the driving quest of Western civilization, would be derived from the original religious vision and by the intellectual tradition preserved and developed within that tradition.

The outward movement of Western civilization in taking possession of the planet began in the ninth and tenth centuries when the

Frankish Empire integrated by Charlemagne began its defense against the incoming powers of the Normans from the north, the Magyars from the east, and the Muslims from the southwest. This movement of resistance turned outward with the Crusades, a movement continued by the later exploration and colonization of the greater part of the entire planet.

Meanwhile the other classical traditions were developing their own wisdom traditions. Immense collections of scriptures and commentaries and expository treatises were compiled through the centuries for understanding the numinous forces moving throughout the universe and their guidance of human affairs in all life circumstances. These traditions governed social life and individual ethical conduct, political authority, family traditions, and childhood formation. They shaped the language of the various societies, their thought patterns, their rituals, their sense of social responsibility, their ruling authority.

This was the wisdom that found expression in the great temples and monuments, the arts, the great literature of the world, as well as in the dances and music traditions. Among these accomplishments we find the rock-hewn temples of India, the shrines and carving of Angkor Wat in Cambodia, the Great Stupa of Borobudur in Indonesia, the Pyramids of Egypt, and of the Mayan and Aztec peoples of Meso-America. So too there appeared the Parthenon of Greece; the Forum, the Pantheon, and the Colosseum of Rome; the Gothic cathedrals of medieval Europe.

Most powerful in shaping the mind of Western civilization were the universities that were founded in the medieval period. Among these were the universities of Paris, Oxford and Cambridge, Prague and Vienna, Salamanca and Bologna. These and other centers of learning have been among the most powerful institutions that have come down into our own times. They provided the context in which later scientific learning could develop and be communicated to our own times.

To list all the achievements of this period in the various civilizations or even in the Western world would be an overwhelming task. These are only fragments of what might be listed to indicate some of the accomplishments of these civilizations that have established the life interpretation of these peoples. Those with any knowledge of this expansive view of the past several thousand years can appreciate the manner in which the contemporary world is still guided by the understanding of the universe and the life interpretation provided by this immense heritage.

When we look back over all these achievements a certain poignant feeling arises in the human soul as we reflect on what had to be endured in Western civilization by the people of these times. A harsh and bitter aspect of the entire period endured by the peasantry was later depicted in T. S. Eliot's play *Murder in the Cathedral,* the story of Thomas à Becket. There was indeed the feeling of participation in some Great Work and yet there was also frequent starvation, minimal protection from the chill of winter, the oppression by those in power, the hardships of women, the severity used against those considered to have deviated from the prescribed beliefs, the needless wars. These gave to the period an ambivalence that it would carry throughout later times when social critics such as Karl Marx and Peter Kropotkin would identify the harshness endured by the populace for the grandeur experienced by the elite.

Yet, however betrayed, throughout all the traditions there were truly noble ideals, enduring insights, valid directions for the human project. There was compassion for the oppressed, the suffering, even the willingness to endure vicarious suffering for others, as with Vimalakirti and Shantideva in the early centuries of Mahayana Buddhism. Ideals of gentleness were expressed in the Taoist traditions of China. Confucianism proclaimed that if the emperor saw someone afflicted on the roadside he should feel responsible for his condition. Chang Tsai in the twelfth century, taught that affection should be shown toward everyone: "Even those who are tired and infirm, crippled or sick, those who have no brothers or children, wives or husbands, are all my brothers who are in distress and have no one to turn to" (quoted in deBary and Lufrano, vol. 2, pp. 683–684). So too in Western tradition are the examples of Francis of

Assisi in his intimacy with the natural world and in his commitment to the poor.

Among the contributions of Western civilization to the human project and to the destinies of the planet might be listed not only the historical realism but also the emphasis on human intelligence. To appreciate this confidence in the powers of the human mind developed in the medieval European world, we need to reflect on the situation in Europe in the twelfth and thirteenth centuries. This was the culminating period in the philosophical and cultural development of Islamic society in Spain. There were situated the great schools of philosophy based on the heritage of Greek thought, especially that of Aristotle. At this time the thought of Aristotle passed over from Islamic Spain into the Christian world of medieval Europe. The crisis concerning faith and reason and how these were related to each other that had arisen in twelfth century Muslim Spain developed also in the Christian world. To deal with this question and to develop an authentic Christian intellectual tradition, Thomas Aquinas was brought to Rome from the University of Paris in 1259. From that time until his death in 1274 he dedicated himself to the interpretation of biblical revelation within the context of the Aristotelian cosmology.

Mainly through his work the Christian world developed a confidence in the reasoning processes of the human mind, accepting the view of Thomas that no authentic truth known by the human mind could be opposed to any revealed truth. This confidence in the powers of human reason in medieval Europe and in the other civilizations is a central feature of the wisdom of the traditions. This commitment in the European world is what made the succeeding age of science possible, although the difference between the mainly deductive reasoning of Aristotle and the mainly empirical inquiry of modern science created a tension that has continued throughout recent centuries.

THE WISDOM of science, as this exists in the Western world at the beginning of the twenty-first century, lies in its discovery that the universe has come into being by a sequence of evolutionary transformations over an immense period of time. Through these transformation episodes the universe has passed from a lesser to a greater complexity in structure and from a lesser to a greater mode of consciousness. We might say that the universe, in the phenomenal order, is self-emergent, self-sustaining, and self-fulfilling. The universe is the only self-referent mode of being in the phenomenal world. Every other being is universe-referent in itself and in its every activity. Awareness that the universe is more cosmogenesis than cosmos might be the greatest change in human consciousness that has taken place since the awakening of the human mind in the Paleolithic Period.

This earlier awakening of the human mind took place in a spatial mode of understanding. The universe, as originally experienced by the human mind, moved in an ever-renewing sequence of changes that easily coordinated with the changes in the natural world, with the daily cycle of dawn and dusk, with the yearly cycle of the seasons. In this context the great journey toward life fulfillment is the journey portrayed in the mandala symbol where the human journey toward fulfillment is toward the center where the divine, the cosmic, and the human worlds become present to each other in mutual fulfillment. The small self of the individual reaches its completion in the Great Self of the universe.

A constant awareness of this spatial context of life gives to human life a deep security, for this ever-renewing world is both an abiding and a sacred world. To live consciously within this sacred world is for the personal self of the individual to be integral with the Great Self of the universe. To move from this abiding spatial context of personal identity to a sense of identity with an emergent universe is a transition that has, even now, not been accomplished in any comprehensive manner by any of the world's spiritual traditions.

This change in human consciousness had its beginning in the sixteenth century with Copernicus. At this time both the value and the difficulty in the work of Thomas Aquinas became apparent.

Copernicus and his followers such as Kepler and Galileo could not have done their work with such confidence unless Thomas had authenticated the reasoning function of the human mind in Western tradition. The difficulty was that Thomas had done his work too well, for he had established Christian revelation so fully within the scientific perspective of Aristotle that it now appeared that any discoveries made that opposed the view of the universe as described by Aristotle must necessarily be false. If they were false then they could not be coherent with revealed teaching, since one of the perspectives of Thomas was that any error concerning the natural world would endanger the authentic understanding of the world of faith.

Because the religious commitment to Aristotle was so intense it was unavoidable that a conflict should occur once such new developments in science began to take place. It was not simply a commitment to the science of Aristotle, it was a commitment also to the deductive processes of reason that tended to dominate all such earlier thinking. Only very late in the history of human thinking did the full appreciation of empirical research science come into being. When it did come into being it reacted with understandable intensity against traditional deductive processes. The reaction was not only to the mode of thinking and to the structure of the world presented, but it was extended even to the denial of the spiritual realm.

For the first time in human history the spirit world, the world of soul, was considered an unreal emotional or aesthetic experience of the human psyche. As a subjective illusion, without acceptable evidence it had no objective validity. Scientists took over both the intellectual and moral guidance of the society through their control over the human mind in the educational program. By its inventive genius science also brought forth an endless number of new technologies that gave to humans amazing power over the phenomenal world.

Francis Bacon, in the early seventeenth century, proposed that through experiments with nature, we could learn more about just how nature functions and through this knowledge we could control nature rather than be controlled by nature. While this was a profound encouragement to the idea of experimentation, it was not Francis Bacon but Galileo Galilei who first performed thoroughly controlled and mathematically measured experiments. His work, together with that of Johannes Kepler, who had first observed that the planets move in elliptical rather than circular orbits, set the background for the work of Isaac Newton who came to understand the laws of gravitation in relation to the movement of the heavenly bodies. His description provided the dominant Western concept of the universe until the time of Albert Einstein and Max Planck in the twentieth century.

Newton, however, had no idea of the evolutionary nature of the universe. This insight came later through continued studies of the universe, but also of the geological structure and biological systems of Earth. These studies eventually led to an awareness that not only Earth but the entire universe had come into being through a long sequence of evolutionary transformations over an immensely long period of time. The important thing about all these discoveries is that they led to an awareness of the unity of the universe within itself and with each of its components. It also led to a realization that each component of the universe is immediately in contact with each of the other components of the universe. In this manner it could be said that in a scientific as well as a religious sense the small self of the individual finds its Great Self in the universe. These somehow exist for each other.

Because this story is a single story and the components of the universe are so intimately related, the story must account for human intelligence. If we consider that human intelligence is a psychic faculty, then the universe from the beginning must be a psychic-producing process. To find a place for the human is the difficulty of those who would maintain that the universe is simply a material mode of being without an intelligent dimension.

If the unity of the universe is one aspect of the wisdom of science, another aspect is the emergent nature of the universe. The third is the existence of human intelligence as an integral component of the universe. The story of the universe becomes the epic story of our times. It narrates something that can be considered in analogy with the epic of the *Odyssey* and with the other epic stories such as the *Mahabharata* and the *Ramayana* of India, or the *Niebelungenlied* of the Germanic world.

AFTER CONSIDERING the wisdom of indigenous peoples, the wisdom of women, the wisdom of the traditions, and the wisdom of science, it seems quite clear that these all agree in the intimacy of humans with the natural world in a single community of existence. The human emerges from the larger universe and discovers itself in this universe. This we find expressed throughout the life and thought and ritual of indigenous peoples. In the wisdom of women it is found in the description of the universe as a mutually nourishing presence of all things with each other. Such is the view of the universe presented in the Goddess figure and other symbolisms. After being excluded from so much of the human world over the centuries, women are revealing the disaster of androcentrism to our society for the first time in Western history.

So too in the classical traditions, the basic teaching in all of the traditions is the fulfillment of the human in the larger functioning of the universe. This we find in Hinduism in the unity of the individual self with the *Atman*, the Great Self of the universe. This is also present in the Buddhist teaching that every mode of being participates in the Buddha Nature. In the Chinese expression of the "One Body" we are told by Wang Yang-ming in the sixteenth century: "Everything from ruler, minister, husband, wife, and friends to mountains, rivers, heavenly and earthly spirits, birds, animals, and plants, all should be truly loved in order to realize my humanity that forms a unity, and then my clear character will be completely manifested, and I will really form one body with Heaven Earth, and the myriad things."

In the Western world the unity of humans with the other components of the universe in a single integral-entity universe finds expression most clearly in the cosmology of Plato as expressed in the *Timeaeus* (par. 36e), "Now when the creator had framed the soul according to his will, he formed within her the corporeal universe, and brought the two together and united them, center to center." Yet it was the Stoic philosopher, Chrysippus, with his idea of the great city of the universe, the *Cosmopolis,* that most clearly expressed this Western sense of oneness of the universe within a single community of being by a political analogy. Throughout the medieval period the unity of humans with the larger universe was founded more on the creation story of Genesis and on the cosmic dimension of the Christ presence in the universe as expressed by Saint Paul in his *Epistle to the Colossians* where he indicates that in the mystical Christ "all things hold together."

A new basis for the unity of humans with the larger earth community is found in the discoveries of modern science. The more clearly we understand the sciences and their perceptions of the universe, the more clearly we appreciate the intimate presence of each component of the universe with every other component. This unity is realized both in our studies of the large-scale structure and functioning of the universe and in the geobiological systems of the Earth.

A similar unity is found in the science traditions of the Western world. The more clearly we understand the sciences and their perceptions of the universe, the more clearly we understand the intimate presence of each component of the universe with every other component. This unity is realized in a unique manner in the geobiological systems of the Earth.

It becomes increasingly evident that in our present situation no one of these traditions is sufficient. We need all of the traditions. Each has its own distinctive achievements, limitations, distortions, its own special contribution toward an integral wisdom tradition that seems to be taking shape in the emerging twenty-first century. Each of the traditional modes of understanding seems to be experiencing a renewal. For the first time the indigenous traditions are accepted as setting the basic model for human presence to the universe. We need such intimacy with the natural world as that presented in the Great Thanksgiving Ceremony of the Iroquois Indians as they made formal recognition of their existence as the gift of the various powers of the universe. The Harvard-based Forum on Religion and Ecology, which grew out of a three-year series of conferences on the world's religions and their views of nature, is an important new direction for examining the wisdom of the religious traditions for guidance into the next century.

For the first time also we begin to understand that the human project belongs in the care and under the direction of both women and men. This was a movement out of a patriarchal society into a truly integral human order. So too the traditional Western civilization must withdraw from its efforts at dominion over the Earth. This will be one of the most severe disciplines in the future, for the Western addiction to economic dominance is even more powerful than the drive toward political dominance.

Then, finally, there is the epic of evolution, the contribution of science toward the future. The universe story is our story, individually and as the human community. In this context we can feel secure in our efforts to fulfill the Great Work before us. The guidance, the inspiration and the energy we need is available. The accomplishment of the Great Work is the task not simply of the human community but of the entire planet Earth. Even beyond Earth, it is the Great Work of the universe itself.

17

MOMENTS *of* GRACE

AS WE ENTER THE TWENTY-FIRST CENTURY, WE ARE EXPERIencing a moment of grace. Such moments are privileged moments. The great transformations of the universe occur at such times. The future is defined in some enduring pattern of its functioning.

There are cosmological and historical moments of grace as well as religious moments of grace. The present is one of those moments of transformation that can be considered as a cosmological, as well as a historical and a religious moment of grace.

Such a moment occurred when the star out of which our solar system was born collapsed in enormous heat, scattering itself as fragments in the vast realms of space. In the center of this star the elements had been forming through a vast period of time until in the final heat of this explosion the hundred-some elements were present. Only then could the sun, our star, give shape to itself by gathering these fragments together with gravitational power and then leaving some nine spherical shapes sailing in elliptical paths around itself as planetary forms. At this moment Earth too could take shape; life could be evoked; intelligence in its human form became possible.

This supernova event of a first or second generation star could be considered a cosmological moment of grace, a moment that determined the future possibilities of the solar system, Earth, and of every form of life that would ever appear on the Earth.

For the more evolved multicellular organic forms of life to appear there then had to appear the first living cell: a procaryotic cell capable, by the energy of the sun, the carbon of the atmosphere, and the hydrogen of the sea, of a metabolic process never known previously. This original moment of transition from the nonliving to the living world, was fostered by the fierce lightning of these early times. Then, at a critical moment in the evolution of the original cell, another cell capable of using the oxygen of the atmosphere with its immense energies appeared. Photosynthesis was completed by respiration.

At this moment the living world as we know it began to flourish until it shaped the Earth anew. Daisies in the meadows, the song of the mockingbird, the graceful movement of dolphins through the sea, all these became possible at this moment. We ourselves became possible. New modes of music, poetry, and painting, all these came into being in new forms against the background of the music and poetry and painting of the celestial forms circling through the heavens.

In human history there have also been such moments of grace. Such was the occasion in northeast Africa some 2.5 million years ago when the first humans stood erect and a cascade of consequences was begun that eventuated in our present mode of being. Whatever talent exists in the human order, whatever genius, whatever capacity for ecstatic joy, whatever physical strength or skill, all this has come to us through these earlier peoples. It was a determining moment.

There were other moments too in the cultural-historical order when the future was determined in some comprehensive and beneficial manner. Such a moment was experienced when humans first were able to control fire; when spoken language was invented; when the first gardens were cultivated; when weaving and the shaping and firing of pottery were practiced; when writing and the alphabet were invented. Then there were the moments when the great visionaries were born who gave to the peoples of the world their unique sense of the sacred, when the great revelations occurred. So too there were the times when the great storytellers appeared—Homer and Valmiki and other composers who gave to the world its great epic tales. There was also the time of the great historians—Ssu-ma Ch'ien in China, Thucydides in Greece, Ibn Khaldun in the Arab world.

So now in this transition period into the twenty-first century, we are experiencing a moment of grace, but a moment in its significance that is different from any previous moment. For the first time the planet is being disturbed by humans in its geological structure and its biological functioning in a manner like the great cosmic forces that alter the geological and biological structures of the planet or like the glaciations.

We are also altering the great classical civilizations as well as the indigenous tribal cultures that have dominated the spiritual and intellectual development of vast numbers of persons throughout these past five thousand years. These civilizations and cultures that have governed our sense of the sacred and established our basic norms of reality and value and designed the life disciplines of the peoples of Earth are terminating a major phase of their historical mission. The teaching and the energy they communicate are unequal to the task of guiding and inspiring the future. They cannot guide the great work that is before us. We will never be able to function without these traditions. But these older traditions alone cannot fulfill the needs of the moment. That they have been unable to prevent and have not yet properly critiqued the present situation is evident. Something new is happening. A new vision and a new energy are coming into being.

After some four centuries of empirical observation and experiment we are having a new experience of the deepest mysteries of the universe. We see the universe both as a developmental sequence of irreversible transformations and as an ever-renewing sequence of seasonal cycles. We find ourselves living both as cosmos and as cosmogenesis. In this context we ourselves have become something of a cosmic force. If formerly we lived in a thoroughly understood ever-renewing sequence of seasonal change, we now see ourselves both as the consequence of a long series of irreversible transformations and as a determining force in the present transformation that the Earth is experiencing.

As happened at the moment when the amount of free oxygen in the atmosphere threatened to rise beyond its proper proportion and so destroy all living beings, so now awesome forces are let loose over the Earth. This time, however, the cause is from an industrial economy that is disturbing the geological structure and life-systems of the planet in a manner and to an extent that the Earth has never known previously. Many of the most elaborate expressions of life and grandeur and beauty that the planet has known are now threatened in their survival. All this is a consequence of human activity.

So severe and so irreversible is this deterioration that we might well believe those who tell us that we have only a brief period in which to reverse the devastation that is settling over the Earth. Only recently has the deep pathos of the Earth situation begun to sink into our consciousness. While we might exult in our scientific and technological achievements in our journey to the moon we must also experience some foreboding lest, through our industrial uses of these same scientific and technological processes, we reduce the wonder and beauty as well as the nourishing capacities of the Earth. We might lose the finest experiences that come to us through all those wondrous forms of life expression as well as the sources of the food and clothing and shelter that we depend on for our survival.

It is tragic to see all those entrancing forms of life expressions imperiled so wantonly, forms that came into being during the past 65 million years, the lyric moment of Earth development. Yet as so often in the past, the catastrophic moments are also creative moments. We come to appreciate the gifts that the Earth has given us.

Such is the context in which we must view this transition period into the twenty-first century as a moment of grace. A unique opportunity arises. For if the challenge is so absolute, the possibilities are equally comprehensive. We have identified the difficulties but also the opportunities of what is before us. A comprehensive change of consciousness is coming over the human community, especially in the industrial nations of the world. For the first time since the industrial age began we have a profound critique of its devastation, a certain withdrawal in dismay at what is happening, along with an enticing view of the possibilities before us.

Much of this is new. Yet all during the last few decades of the twentieth century studies were made that give us precise information on what we must do. A long list of persons, projects, institutions, research programs, and publications could be drawn up indicating that something vital is happening. A younger generation is growing up with greater awareness of the need for a mutually enhancing mode of human presence to the Earth. We have even been told that concern for the environment must become "the central organizing principle of civilization" (Gore, p. 269).

The story of the universe is now being told as the epic story of evolution by scientists. We begin to understand our human identity with all the other modes of existence that constitute with us the single universe community. The one story includes us all. We are, everyone, cousins to one another. Every being is intimately present to and immediately influencing every other being.

We see quite clearly that what happens to the nonhuman happens to the human. What happens to the outer world happens to the inner world. If the outer world is diminished in its grandeur then the emotional, imaginative, intellectual, and spiritual life of the human is diminished or extinguished. Without the soaring birds, the great forests, the sounds and coloration of the insects, the free-flowing streams, the flowering fields, the sight of the clouds by day and the stars at night, we become impoverished in all that makes us human.

There is now developing a profound mystique of the natural world. Beyond the technical comprehension of what is happening and the directions in which we need to change, we now experience the deep mysteries of existence through the wonders of the world about us. This experience has been considerably advanced through the writings of natural-history essayists. Our full entrancement with various natural phenomena is presented with the literary skill and interpretative depth appropriate to the subject. We experience this especially in the writings of Loren Eiseley, who recovered for us in this century the full wonder of the natural world about us. He has continued the vision of the universe as this was presented to us in the nineteenth century by Ralph Waldo Emerson, Henry David Thoreau, Emily Dickinson, and John Muir.

We are now experiencing a moment of significance far beyond what any of us can imagine. What can be said is that the foundations of a new historical period, the Ecozoic Era, have been established in every realm of human affairs. The mythic vision has been set into place. The distorted dream of an industrial technological paradise is being replaced by the more viable dream of a mutually enhancing human presence within an ever-renewing organic-based Earth community. The dream drives the action. In the larger cultural context the dream becomes the myth that both guides and drives the action.

But even as we make our transition into this new century we must note that moments of grace are transient moments. The transformation must take place within a brief period. Otherwise it is gone forever. In the immense story of the universe, that so many of these dangerous moments have been navigated successfully is some indication that the universe is for us rather than against us. We need only summon these forces to our support in order to succeed. Although the human challenge to these purposes must never be underestimated, it is difficult to believe that the larger purposes of the universe or of the planet Earth will ultimately be thwarted.

Bibliography

THIS BIBLIOGRAPHY INCLUDES A WIDE RANGE OF SOURCE MATERIALS. Such extensive materials are required because of the comprehensive nature of the issues dealt with in this study. Here we are dealing with the wide range of human affairs in relation to the planet Earth. I have given a brief explanation of each item so that the reader will be acquainted with the issues presented. I have included writings from points of view diverse from my own. While the greater number of books are readily available, some were published earlier and may be difficult to locate. Yet these books are of special value for their contribution to the subject under discussion. Still, as long as this bibliography is, it is entirely inadequate in relation to the immense amount of material that deserves to be included. If some of the best materials are missing, I can only say that the books listed can at least be taken as a beginning.

Abram, David. *The Spell of the Sensuous: Perception and Language in a More-Than-Human World.* New York: Pantheon Books, 1996. A presentation of the manner in which humans symbolize their experience of the world about them, with special attention to what happens when this symbolization finds expression in language and in writing. Written with a new depth of insight into human-Earth relations.

Allen, Paula Gunn. *The Sacred Hoop: Recovering the Feminine in American Indian Tradition.* Boston: Beacon Press, 1986. A scholar of American Indian history and literature, a poet, and an essayist. Allen, with her Laguna Pueblo heritage, presents a rich and varied collection of essays dealing largely with the role of the feminine in the life and literature of the indigenous peoples of the North American continent.

Anderson, Robert O. *Fundamentals of the Petroleum Industry.* Norman, OK: University of Oklahoma Press, 1984. A valuable survey of the petroleum industry in its many aspects. Much of this data is not readily available elsewhere.

Anderson, William. *Green Man: The Archetype of Our Oneness with the Earth.* Photography by Clyde Hicks. San Francisco: HarperCollins, 1990. That the human had a certain identity with the Earth finds expression in the artistic form throughout Western history. The work of medieval artists is especially impressive.

Aquinas, Thomas. *Summa Contra Gentiles.* Translated by Anton C. Pegis. Notre Dame, IN: University of Notre Dame Press, 1955. The most significant work of Thomas Aquinas, written for those who do not accept the Christian scriptures. (Abbreviated as SCG.) References are cited as the book and the appropriate chapter. There are four books in this work; the chapters are the equivalent of extended paragraphs in contemporary writings.

———. *Summa Theologica.* Translated by English Dominicans. New York: Benziger Brothers, 1946. The thirteenth-century masterwork of Thomas Aquinas explaining Christian belief within the context of Aristotelian cosmology. (Abbreviated as ST.) The work is written in terms of questions proposed and responses given. References are cited as one of the three parts of the work, then the number of the question, then the number of the article answering the question.

Ayers, Harvard, et al., eds. *An Appalachian Tragedy: Air Pollution and Tree Death in the Eastern Forests of North America.* San Francisco: Sierra Club Books, 1998. The text of this photographic volume is by Charles Little, someone with a comprehensive acquaintance with the forests of the North American continent. The photographs are rendered with the perfection that is now possible.

Bailey, Ronald, ed. *The True State of the Planet: Ten of the World's Premier Environmental Researchers in a Major Challenge to the Environmental Movement.* New York: The Free Press, 1995. A work by ten scholars in various fields of expertise, all claiming that the industrial venture is not ruining the planet but is rather assisting efforts to maintain the well-being of the planet, this is a direct assault on the environmental movement, sponsored by the Competitive Enterprise Institute. It is of value to know how prevalent the antagonism to the environmental movement is.

Baring, Anne, and Jules Cashford. *The Myth of the Goddess: Evolution of an Image.* London and New York: Penguin Books, 1991. A monumental survey of Goddess worship in various cultural traditions with abundant literary references and impressive illustrations. Written with clarity and grace of expression.

Barlow, Connie, ed. *Evolution Extended: Biological Debates on the Meaning of Life*. Cambridge: MIT Press, 1994. An extensive collection and interpretation of data on recent biological studies concerning the evolutionary story and the meaning that it has for the human story. Barlow is a writer and editor with both scientific insight and expository skills.

Barnet, Richard J., and John Cavanaugh. *Dreams: Imperial Corporations and the New World Order.* New York: Simon and Schuster/Touchstone Books, 1994. A vast amount of information on the functioning of the corporations, their control of human affairs, and the consequences in every phase of human social and cultural development.

Barney, Gerald O., ed. *The Global 2000 Report to the President, Entering the Twenty-first Century.* "Commissioned by Carter, Disregarded by Reagan, Published Here in an Unabridged Edition." New York: Penguin Books, 1982. This volume has special significance as the first effort sponsored by the American political establishment to make a comprehensive survey of the economic-political problems that will surely arise as the industrial order finds itself in a world of ever-diminishing natural resources.

Berger, John J. *Restoring the Earth: How Americans Are Working to Renew Our Damaged Environment.* New York: Doubleday, 1987. An impressive series of grassroots programs for restoring the Earth are presented here in full detail.

Berry, Wendell. *The Unsettling of America: Culture and Agriculture.* New York: Weatherhill Press, 1986. A classic treatise by an active American farmer on the relationship between the culture of a people and the cultivation of the land. Berry describes how all the basic values in this way of life become distorted by industrial agriculture.

Bertell, Rosalie. *No Immediate Danger: Prognosis for a Radioactive Earth.* Toronto, Canada: Women's Educational Press, 1985. A careful and detailed examination of the consequences of nuclear radiation on the biosystems of the natural world, with special attention to the consequences on the human population.

Billington, Ray Allen. *Land of Savagery, Land of Promise: The European Image of the American Frontier in the Nineteenth Century.* New York: W. W. Norton, 1981. An impressive collection of data from the writings of

Europeans, mostly of the nineteenth century, but with extensive data from the eighteenth. Author is concerned with the various interpretative contexts of the writers, from romantic idealism to extreme realism.

Boff, Leonardo. *Cry of the Earth, Cry of the Poor.* Maryknoll, NY: Orbis Books, 1997. An impressive insight into the relation between the social issue and the ecology issue from a Latin American liberation theologian with intimate experience of both issues. He shows how neither ecological improvement nor social well-being will function separate from each other.

Bohm, David. *Wholeness and the Implicate Order.* London: Routledge and Kegan Paul, 1980. This philosopher has given his attention especially to the immediate relationship between the smallest particle and the larger community of existence. Bohm provides a way of understanding how the universe can be integral with itself throughout its vast extent in space and its sequence of transformations in time.

Bookchin, Murray. *The Ecology of Freedom: The Emergence and Dissolution of Hierarchy.* Palo Alto, CA: Cheshire Books, 1982.

———. *Remaking Society.* Montreal: Black Rose Books, 1989. The earliest of our American social ecologists, Murray Bookchin is profoundly dedicated to the elimination of hierarchical controls in human societies. He is also concerned with the harm done by the dominion of humans over nature.

Bourdon, David. *Designing the Earth: The Human Impulse to Shape Nature.* New York: Harry N. Abrams Press, 1995. A presentation of the large-scale work of humans in monumental architectural structures imposed on the Earth and in the carving of the Earth itself.

Bowers, C. A. *The Culture of Denial: Why the Environmental Movement Needs a Strategy for Reforming Universities and Public Schools.* Albany, NY: State University of New York Press, 1997. One of the most competent educators outlines what must be done to evoke a sense of the universe and the role of the human in the universe within educational programs at all levels.

———. *Educating for an Ecologically Sustainable Culture: Rethinking Moral Education, Creativity, Intelligence, and Other Modern Questions.* Albany, NY: State University of New York Press, 1995. An extensive consideration of the deeper cultural forces that brought about the ecological crisis and the cultural changes needed to remedy our present situation, with special attention to education.

Breton, Mary Joy. *Women Pioneers for the Environment.* Boston: Northeastern University Press, 1998. A book that deserves reading by everyone with concern for both the role of women in our society and for healing the ecological disruption caused by modern industry. Until recently men have been better known in nature writing, women have been doing the work of revealing the pollution taking place and working to stop the devastation of the natural world. More than forty of these women are identified here.

Brock, William H. *The Norton History of Chemistry.* New York: W. W. Norton, 1992. A reliable and well-written survey of the long sequence of developments that have led to our present competence in the field of chemistry. Since an immense volume of chemical toxins is disturbing the life-systems of the planet, some acquaintance with the backgrounds of the chemical industry is helpful.

Brower, David, with Steve Chapple. *Let the Mountains Talk, Let the Rivers Run: A Call to Those Who Would Save the Earth.* New York: HarperCollinsWest, 1995. David Brower is a person of rare dedication and efficacy in his leadership in guiding environmental organizations and saving the natural regions of North America. He is perhaps the person in these times to be considered as successor to John Muir or Henry Thoreau. This is a book of reflections in his later years with all the insight and inspiration that we are accustomed to receiving from him.

Brown, Lester R., and Hal Kane. *Full House: Reassessing the Earth's Population Carrying Capacity.* New York: W. W. Norton WorldWatch Series, 1994. A well-documented account of the projected population increase and expected food consumption in relation to the resources for growing the food needed for an acceptable level of survival. Brown has been the moving power in the WorldWatch Institute and in the annual publication of *The State of the World.*

Brown, Lester, et al., eds. *Vital Signs.* New York: W. W. Norton. An annual publication of the WorldWatch Institute since 1992. A survey of the health of the Earth in its basic components, such as air, water, and food supply.

Buell, Lawrence. *The Environmental Imagination: Thoreau, Nature Writing, Nature Writing and the Formation of American Culture.* Cambridge: Harvard University Press, 1995. The environmental consciousness has a long history of literary expression in poetry and especially in the natural history essay. This study of environmental consciousness is an important contribution to understanding our contemporary American culture.

Burdick, Donald L., and William L. Leffler. *Petrochemicals in Nontechnical Language.* Tulsa, OK: PennWell Publishing, 1990. A clear and authentic presentation of the petrochemical industry, one of the central industries in its immediate effect on the planet Earth and its air, water, soil, and biological organisms.

Burger, Julian. *The Gaia Atlas of First Peoples: A Future for the Indigenous World.* New York: Doubleday Anchor Books, 1990. An extremely useful comprehensive survey of the tribal peoples of the world in all their diversity, with basic information on their present status.

Cajete, Gregory. *Look to the Mountain.* Durango, CO.: Kivaki Press, 1994. A Pueblo Indian perspective on Native American worldviews and ecology.

Callicott, J. Baird. *Earth's Insights: A Multicultural Survey of Ecological Ethics from the Mediterranean Basin to the Australian Outback.* Berkeley: University of California Press, 1994. Baird Callicott has long been dedicated to the basic cultural need for a land ethic supporting an intimate rapport with the natural world about us. Here he surveys many of the world's religious traditions in support of environmental ethics.

Campbell, Colin J. *The Coming Oil Crisis.* Brentwood, Essex, England: Multiscience Publishing and Petroconsultants, 1997. The author has followed the petroleum industry in minute detail throughout his professional life. He documents: the discovery, the amount of petroleum available, the rate of use, with careful mathematical measurements as regards its future availability. He has also consulted other authorities in this field before making the final calculations recorded here.

Carson, Rachel. *The Sense of Wonder.* Photographs by Charles Pratt and others. New York: Harper and Row, 1956. Here we are presented with a profound appreciation of the awakening of wonder within children at their first experiences of the natural world and the wonderful discoveries that can be made, especially at the seashore.

———. *Silent Spring.* Boston: Houghton Mifflin, 1962. This is the work of a research biologist sensitive to the consequences of the use of chemicals in efforts to suppress pests and weeds in agriculture, with special reference to the damage done by DDT. This study evoked an intense opposition to Carson's work in scientific circles, in the media and, of course, in the industrial world.

Chung, Hyun Kyung. *Struggle to Be the Sun Again: Introducing Asian Women's Theology.* Maryknoll, NY: Orbis Books, 1990. This book is written with superb insight into the inherent nobility of women amid the oppressions that women have endured. Some sense of the grace and understanding of the author's writing can be seen in her summary statement that theology for Asian women is a language of "hope, dreams, and poetry."

Clinebell, Howard. *Ecotherapy: Healing Ourselves, Healing the Earth.* Minneapolis, MN: Fortress Press, 1996. The author is one of the most competent of contemporary writers on the healing relations between the individual, society, and the natural world.

Cobb, John B., Jr. *Sustainability: Economics, Ecology and Justice.* Maryknoll, NY: Orbis Books, 1992. A foremost theologian who is also knowledgeable in the field of economics brings these disciplines into their proper relation with ecology.

Colburn, Theo, et al. *Our Stolen Future.* New York: E. P. Dutton, 1996. A detailed and extensively researched study of the insidious influence of chemicals on the biological development and functioning of humans, especially during the period when the fetus is in the womb as well as when children in their early years are most vulnerable.

Cole, Nancy, and P. J. Skerrett. *Renewables Are Ready: People Creating Renewable Energy Solutions.* White River Junction, NY: Chelsea Green Publishing, 1995. A proposal sponsored by the Union of Concerned Scientists that the technologies for using renewable energies including photovoltaics, solar heating devices, wind and hydroelectric turbines, and biomass generators, are now available for shifting from nonsustainable fossil fuel energies to sustainable energies.

Costanza, Robert, et al. *An Introduction to Ecological Economics.* A publication of the International Society for Ecological Economics. Boca Raton, FL: St. Lucie Press, 1997. This author, together with Herman Daly and Richard Norgaard, has been a leading influence in introducing the ecological dimension into the study and application of economics. This is a first volume available for institutions of economics and business administration.

Crèvecœur, J. Hector St. John de. *Letters from an American Farmer.* New York: E. P. Dutton, 1957. First published in 1782. "Letters" here is a literary form for expressing the personal experience of the author in farming in the eastern region of the North American continent in the eighteenth century.

Critchfield, Richard. *Villages.* New York: Doubleday Anchor Books, 1981. A survey of village life based on personal experience of villages throughout Asia, Africa, South America, Mexico. Since humans arose into their more developed modes of cultural expression in the village context, we are, in a manner, village beings. The village remains our proper context. Both a deterioration and a renewal of village life is under way throughout the world.

Cronon, William, Jr. *Changes in the Land: Indians, Colonists, and the Ecology in New England.* New York: Hill and Wang, 1983. This story of New England based on the geography and the ecology of the land has, in my view, established what might be considered as a new genre of history, a history based on the integral relation of human communities with the larger community of the natural world. A valuable addition to the usual political, economic, social, and religiously based historical interpretations.

Cronyn, George W., ed. *American Indian Poetry: An Anthology of Songs and Chants.* New York: Liveright, 1934. An excellent collection of Indian poetry with ritual chants. The introduction is by Mary Austin, a well-known native fiction writer of the Southwest.

Crosby, Alfred W. *Ecological Imperialism: The Biological Expansion of Europe 900–1900.* New York: Cambridge University Press, 1986. It is a startling realization that the biological species that have been transplanted from Europe to other regions of the world have been so severe a disruption of the native biosystems.

Daily, Gretchen C., ed. *Nature's Services: Societal Dependence on Natural Ecosystems.* Washington, D.C., and Covelo, CA: Island Press, 1997. A collection of essays on how humans benefit from the integral functioning of the natural world and the corresponding impasse that occurs to the human venture when these benefits are diminished or eliminated.

Daly, Herman E. *Steady State Economics.* San Francisco: W. H. Freeman Press, 1977. A book that should never be forgotten for its significance in introducing a new period in our understanding of economics. Daly's work succeeded that of Nickolaus Georgescu-Roegan. With their work we begin to understand that any human economy will always be a subsystem of the Earth economy.

Daly, Herman E., and John B. Cobb, Jr. *For the Common Good.* Boston: Beacon Press, 1989. A significant study of the present by a theologian and an economist concerned with both the social well-being in the human world and ecological well-being in the natural world.

Dawkins, Richard. *The Blind Watchmaker: Why the Evidence of Evolution Reveals a Universe Without Design.* New York: W. W. Norton, 1987. A highly praised presentation of evolution considered from a rigorous empirical point of view.

Dawson, Christopher. *The Making of Europe: An Introduction to the History of European Unity.* New York: World Publishing, 1932. A classic in the cultural history of the European world that has kept its value over the years. Required reading for anyone who wishes to understand the cultural experience out of which both Europe and America were born and which still determines the deeper structures of Western civilization.

deBary, Wm. Theodore, Richard Lufrano, and Irene Bloom, eds. *Sources of Chinese Tradition.* Two volumes. New York: Columbia University Press, 1999. An indispensable collection of basic sources in Chinese thought and literature from earliest times until the late twentieth century. First published in 1959, this new edition greatly expands the first edition.

Deloria, Vine. *For This Land: Writings on Religion in America.* New York: Routledge, 1998. A collection of writings done over the past three decades by a foremost American Indian intellectual and cultural critic.

DeMaillie, Raymond J., ed. *The Sixth Grandfather: Black Elk's Teachings Given to John G. Neihardt.* Lincoln, NB: University of Nebraska Press, 1984. This is a more recent study of the notes of John Neihardt, which he took from the verbal narrative of Black Elk as he narrated his life story in the 1930s. DeMaillie is an accomplished and recognized scholar of the Plains Indians.

Devall, William, and George Sessions. *Deep Ecology: Living as if Earth Matters.* Salt Lake City, UT: Peregrine Smith Press, 1985. These two authors are the principal scholars in America committed to the teachings of Arnie Naess, who asserts the continuity between humans and other life-forms in a single interrelated community.

Dickason, Olive Patricia, ed. *The Native Imprint: The Contribution of First Peoples to Canada's Character.* Vol. 1. Alberta, Canada: Athabasca University Press, 1995. A basic source for understanding the native influence on the European culture that moved into North America, from the years of the original discoveries until 1815, the date of the conclusion of the Napoleonic Wars and the convening of the Congress of Vienna, which had consequences especially in the French-Canadian world.

Douglas, William O. *A Wilderness Bill of Rights.* Boston: Little, Brown, 1965. One of the few books written in defense of the inherent rights of natural modes of being, claiming nature should not be degraded simply for utilitarian purposes by humans. Douglas was an Associate Justice of the Supreme Court of the United States from 1939 until 1975. He was also a naturalist of some competence.

Dowie, Mark. *Losing Ground: American Environmentalism at the Close of the Twentieth Century.* Cambridge: MIT Press, 1995. A supporter and yet a severe critic of the functioning of the environmental movement. He is especially critical of the larger environmental organizations for lack of social emphasis, depending too much on official political processes and accepting too much support from the foundations and sources that are themselves causing the environmental disruptions. He wants the environmental movement to be more a social movement and less an ecology movement.

Drucker, Peter E. *Innovation and Entrepreneurship: Practice and Principles.* New York: Harper and Row, 1985. This is one of the more influential writings of the dean, and to some extent the founder, of humanistic management as is widely taught in the colleges and universities of America. His sense of management, for all its deficiencies, was a decided improvement on the managerial style of Frederick Winslow Taylor, based on the engineering model, with its insistence on elaborate time studies in evaluating the efficacy of any production process or the value of any employee.

Dubos, René. *Celebrations of Life.* New York: McGraw Hill, 1981. A French research physician and biologist, whose primary study was soil biology, Dubos came to America early in his professional life. He taught at Rockefeller University, assisted in developing drugs for fighting bacterial infections, and isolated the antibiotic that formed the basis for future chemotherapy. He wrote extensively on ecological issues.

Earley, Jay. *Transforming Human Culture: Social Evolution and the Planetary Crisis.* Albany, NY: State University of New York Press, 1997. The author, a careful and learned scholar in the history of social and cultural evolution, proposes that we must take charge of history and the evolutionary process. We can do this now, he proposes, in the general pattern outlined earlier by Duane Elgin and Ken Wilber. The basic emphasis is on attaining a new level of consciousness.

Ehrenfeld, David. *The Arrogance of Humanism.* New York: Oxford University Press, 1981. The thesis of this book indicates that a basic flaw in Western civilization leads to the devastation of the natural world and the incompetence of religious establishments in dealing with the ecological issues. Arrogance must be listed along with androcentrism and patriarchy as a basic flaw in Western civilization; perhaps it is both a cause and a consequence of these other oppressive attitudes.

Ehrlich, Paul R., and John P. Holdren, eds. *The Cassandra Conference: Resources and the Human Predicament.* College Station, TX: Texas A&M University Press, 1988. A conference of scientists called to discuss the human predicament, it was named Cassandra because of its concern for the devastating future awaiting the human community if the assault on the integral functioning of the natural world continues.

Ehrlich, Paul and Anne. *The Population Explosion.* New York: Simon and Schuster, 1991. Paul and Anne Ehrlich are among the earliest, most perceptive, and most consistent observers of the inherent difficulties that would inevitably follow in both the human community and the natural world from unlimited population increase.

Eiseley, Loren. *The Unexpected Universe.* New York: Harcourt Brace Jovanovich, 1969. Dean of the school of anthropology at the University of Pennsylvania and member of the American Academy of Literature, this author was among the most profound of American naturalists in the twentieth century. He wrote extensively in both prose and poetry.

———. *The Night Country.* New York: Charles Scribner's Sons, 1971. Here we have a confrontation with the darker sources at work in the human psyche and in the social process. The value of this author's work is precisely in his clarity in writing about the dark and the daylight worlds in which our human destinies are worked out.

Eisler, Riane. *The Chalice and the Blade: Our History, Our Future.* San Francisco: Harper San Francisco, 1987, 1995. A full and forceful presentation of the early Goddess culture of the Neolithic Period and the rise of patriarchy in association with the rise of the god cultures in the beginning period of the classical civilizations.

Eldredge, Niles. *Reinventing Darwin: The Great Debate at the High Table of Evolutionary Theory.* New York: John Wiley & Sons, 1995. Evolutionary explanation has changed considerably since Darwin published his book in 1859. To understand just how we arrived at our present explanation, this is the book to read.

———. *Life in the Balance: Humanity and the Biodiversity Crisis.* Princeton, NJ: Princeton University Press, 1998. An important treatment of the role of humans in the current sixth extinction period, along with suggestions for how to halt this current loss of biodiversity.

———. *The Pattern of Evolution.* New York: W. H. Freeman, 1999. A further study of evolution and the pattern in its sequence of developments.

Ellul, Jacques. *The Technological Society.* New York: Alfred Knopf/Vintage Books, 1964. Original French edition, 1954. For a critique of modern industrial technologies and their inherent deleterious influence on the human and spiritual cultures of the Western world this presentation remains unique for the depth of its analysis of what has happened in the twentieth century and the profound consequences of technological civilization on the human mode of being.

Fagin, Dan, Marianne Lavelle, and the Center for Public Integrity. *Toxic Deception: How the Chemical Industry Manipulates Science, Bends the Law, and Endangers Your Health.* Secaucus, NJ: Birch Lane Press, 1996. A comprehensive study of the commercial imposition of a multitude of toxic chemicals on the American public.

Flannery, Timothy Fridtjof. *Future Eaters: An Ecological History of the Australasian Lands and People.* New York: George Braziller, 1994. A unique study of the species extinction that occurred under the influence of earlier peoples based on thorough research and careful interpretation of the influence on other species by the early human inhabitants of Australasia. It is something of a revelation to those who consider that indigenous peoples are consistently benign in their relation with other forms of life on Earth.

Fletcher, W. Wendell, and Charles E. Little. *The American Cropland Crisis: Why U.S. Farmland Is Being Lost and How Citizens and Governments Are Trying to Save What Is Left.* Bethesda, MD: American Land Forum, 1982. A useful study of the need to appreciate the agricultural land of the North American continent as the greatest body of cropland on the planet. The lack of any effective program to preserve this land in its integrity is a profound failure of the administration, the Congress, and the people. The study presented here is one of the finest statements of what needs to be done.

Frankfort, Henri, et al. *Before Philosophy: The Intellectual Adventure of Ancient Man.* Baltimore: Penguin, 1949. (Originally published by University of Chicago Press in 1946.) An archaeologist with extensive experience in the ancient Near East, the author had a valuable insight into the coordination of human affairs with the structure and functioning of the larger cosmological order.

Frieden, Bernard J. *The Environment Protection Hustle.* Cambridge: MIT Press, 1979. A critique of the environmental movement as elitist and without concern for the real problems of the people. This study is based on research done in response to the efforts to limit development in California in the 1970s.

Friedman, Lawrence M. *A History of American Law.* 2d ed. New York: Simon and Schuster, 1985. A thorough and quite readable account of the development of law and its manner of functioning in America. An understanding of the legal context of human relations with the natural world is necessary for the success of any environmental activity. This should be read in relation to the work by Morton Horowitz that is also cited in this bibliography.

Fumento, Michael. *Science Under Siege: How the Environmental Misinformation Compaign Is Affecting Our Lives.* New York: William Morrow, 1993. An emotionally based assault on ecological writings as being scientifically untenable.

Fuson, Robert H. *The Log of Christopher Columbus.* Camden, ME: International Marine Publishing Company, 1987. A more recent translation of the earliest of our sources for the first impressions of Europeans on establishing communication with the indigenous peoples of the Americas.

Gever, John, et al. *Beyond Oil: The Threat to Food and Fuel in the Coming Decades.* Cambridge: Ballinger Publishing Company, 1986. A valuable study of the consequences of the depletion of petroleum on agricultural production due to a decline in the availability of oil-based fertilizer.

Gimbutas, Marija. *The Language of the Goddess.* San Francisco: Harper San Francisco, 1989. A scholar of some renown in her archaeological inquiry into the mythology of the Goddess in the late Paleolithic and throughout the Neolithic Periods of human development, in the region of Asia Minor and the Balkan region of what is considered Old Europe.

Glacken, Clarence J. *Traces on the Rhodian Shore: Nature and Culture in Western Thought: From Ancient Times to the End of the Eighteenth Century.* Berkeley: University of California Press, 1967. A study of the history of Western civilization with an impressive depth of understanding of the role of the human in the larger context of the natural world as this has been envisaged in the various periods of Western development.

Goldsmith, Edward. *The Way: An Ecological World View.* Boston: Shambhala, 1993. One of the most perceptive of contemporary writers on ecological issues, Goldsmith presents a comprehensive statement of the most urgent issues of the present.

Goodenough, Ursula. *The Sacred Depths of Nature.* New York: Oxford University Press, 1998. We could never have guessed that scientific inquiry would open into such a world of wonder for the mind and such joy of heart.

Gore, Albert. *Earth in the Balance.* New York: Houghton Mifflin, 1992. An excellent presentation of the basic ecology issues with sound guidance for a way of dealing with them.

Griffin, Susan. *Woman and Nature.* New York: Harper and Row, 1978. An impressive scholar and writer of the feminist movement expresses her concern over an exaggerated identification of woman with nature.

Grim, John, *The Shaman.* Norman, OK: University of Oklahoma Press, 1984. An important study of the shaman in the context of the typology of religious personalities. Special attention is given to these healing practitioners among the Ojibway peoples.

Hawken, Paul. *The Ecology of Commerce.* New York: HarperCollins, 1993. The author shows a realistic understanding of the ecological crisis and a corresponding awareness of the present commercial-industrial world and its mode of functioning. His hope is that the corporations controlling the present Earth situation will lead us into restorative economics, an economics more sustainable in the future than our present plundering economics.

Hayden, Tom. *The Lost Gospel of the Earth.* San Francisco: Sierra Club Books, 1996. A book of personal experiences and reflections from someone with extensive experience in the world of political affairs. Tom Hayden brings together views of various religions for an ecologically sustainable future.

Helvarg, David. *The War Against the Greens: The Wise-Use Movement, the New Right, and Anti-Environmental Violence.* San Francisco: Sierra Club Books, 1994. A valuable resource for understanding the assault against the environmentalist movement.

Herlihy, David. *The Black Death and the Transformation of the West.* Edited and with an introduction by Samuel K. Cohn, Jr. Cambridge, MA, and London, England: Harvard University Press, 1997. One of the best studies of the Black Death and its consequences in the economic, social, and cultural development of this critical century in Western history.

Herman, Arthur. *The Idea of Decline in Western History.* New York: The Free Press, 1997. A long and detailed overview of the pessimism of the last two centuries concerning the future of civilization just when the doctrine of progress seemed to be realized in the industrial-commercial dominion over nature.

Hessel, Dieter, and Rosemary Radford Ruether, eds. *Christianity and Ecology.* Cambridge: Center for the Study of World Religions and Harvard University Press, 1999. This is the most comprehensive collection of papers to date on the role and resources of Christianity in relation to the environment.

Horowitz, Morton J. *The Transformation of American Law, 1780–1860.* New York: Oxford University Press, 1995. Originally published in 1977. An invaluable work for understanding the bonding of the legal and the judiciary professions in America with the commercial-industrial establishment to the neglect of the ordinary citizen, worker, farmer, and those less affluent or influential.

———. *The Transformation of American Law, 1860–1920.* Cambridge: Harvard University Press, 1994. In this second volume on American law, the most valuable contribution is the description of the rise of the commercial, industrial, and financial corporations, their legal status, and the power they have attained.

Hughes, Robert. *American Visions: The Epic History of Art in America.* New York: Alfred A. Knopf, 1997. Since much of this book is concerned with the relation of the human community with the natural world, it is of great value to see how this relationship is depicted in the various arts. Here the visual arts are presented.

Hunt, Charles B. *Natural Regions of the United States and Canada.* San

Francisco: W. H. Freeman Press, 1974. While this is an older work, it is still valuable for its identifications and thorough descriptions of the various geographical regions of the North American continent.

Huntington, Ellsworth. *Climate and Civilization.* New Haven: Yale University Press, 1915. This is one of the earliest and most thorough studies of the effects of geographical environment on human cultural development.

Hyams, Edward. *Soil and Civilization.* New York: Harper and Row Torchbooks, 1952. A unique book, written in superb English style and with comprehensive erudition, it provides insight into the various civilizations of the past, the manner in which they have dealt with their land, and the consequences on the human community itself. The decline of civilizations seems to be closely related to the lack of care for and the consequent deterioration of their soil.

Irland, Lloyd C. *Wildland and Woodlands: The Story of New England's Forests.* Hanover, NH: University Press of New England, 1982. An excellent description of the forests of New England, their history, their present status, and their future prospects, by someone who is himself a forester with extensive academic training and official experience. He outlines the necessary steps that need to be taken if these forests are to continue in any integral manner.

Jackson, Wes. *Becoming Native to This Place.* Lexington, KY: University Press of Kentucky, 1994. This series of six essays are an expansion of a lecture given at the University of Kentucky on the need for a sense of place, how this was lost and how it can be recovered. The small community in intimate contact with the land is the needed context for a recovery of this sense of place.

———. *New Roots for Agriculture.* New York: Friends of the Earth, 1980. The author, founder and director of the Land Institute in Salina, Kansas, studies the deeper forces at work in the biosystems of the natural world that we need to understand if we are to sustain the health of our agricultural products over a prolonged period of time. He is concerned with the integral health of these sources on which the food supply of humans depends. He is also deeply committed to the permaculture practices of agriculture.

Jansson, AnnMari, et al., eds. *Investing in Natural Capital: The Ecological Economics Approach to Sustainability.* Washington, D.C.: The Free Press, 1994. We need to protect and foster whatever natural capital survives the abuse we have shown to those very forces on which we depend for continued well-being of the human community.

Jantsch, Erich. *The Self-Organizing Universe: Scientific and Human Implications of the Emerging Paradigm of Evolution.* New York: Pergamon, 1980. This is an authentic presentation of the thought of Ilya Prigogine, whose basic study has been in chemistry. Even in the pre-living world an active self-organizing dynamics is functioning. The presentation here includes various levels of self-organizing, from the earliest shaping of matter to the activity of humans. This spontaneous appearance of ordered complexity is one of the main concerns of thoughtful scientists at the present time.

Jensen, Derrick, et al. *Railroads and Clearcuts.* Spokane, WA: Keokee Inland Island Public Lands Council, 1995. Story of the vast timberlands of the Northwest given by Congress in 1864 to the Northern Pacific Railroad to be sold as a commodity, mostly to the Weyerhauser corporate empire, for clearcutting— and the consequences into the present.

Johnson, Hugh. *The International Book of Trees.* New York: Simon and Schuster, 1973. The author has a rare gift for understanding and writing about the various species of trees. He not only describes the various species in technical language, he also communicates something of that deeper reality that is the poetry or the mystique of the various species.

Kauffman, Stuart. *At Home in the Universe: The Search for Laws of Self-Organization and Complexity.* New York: Oxford University Press, 1995. Written by a widely recognized scientist concerned with the comprehensive study of the universe and the source of its order from its physics, through its chemical, biological, and historical human phases. He is a scientist who is also a writer of impressive competence. This is an excellent source to see how thinking scientists are interpreting their own discoveries.

Kay, Jane Holtz. *Asphalt Nation: How the Automobile Took Over America and How We Can Take It Back.* New York: Crown Publishers, 1997. A book long needed to understand just what is happening to the human community as the automobile becomes the central economic feature of the society, with corresponding cultural consequences.

Kaza, Stephanie. *The Attentive Heart: Conversations with Trees.* New York: Ballantine Books, 1993. While others write about intimacy with nature, this author expresses it.

Keller, Evelyn Fox. *A Feeling for the Organism: The Life and Work of Barbara*

McClintock. New York: W. H. Freeman, 1983. One of the most significant books to indicate the value of the personal rapport of the biological scientist with the subject being studied.

Kelley, Kevin W. *The Home Planet.* New York: Addison-Wesley, 1988. A book of photographs chosen from the NASA collection, taken by the astronauts and published with brilliant comments by the astronauts themselves. The poetic and emotional tone of the comments makes a powerful impact.

Kimbrell, Andrew. *The Human Body Shop: The Engineering and Marketing of Life.* San Francisco: Harper San Francisco, 1993. A detailed study of what is happening now in transplanting and modifications we are imposing on the human body. Written by someone with exceptional clarity in his thinking and precision in his writing.

Kolodny, Annette. *The Lay of the Land: Metaphor as Experience and History in American Life and Letters.* Chapel Hill, NC: University of North Carolina Press, 1975. A study of the feminine metaphor for understanding and relating to the New World discovered here in North America. Provides the basis for a critical insight into the forces in American history and culture that have been leading the society into a destructive relation with its land.

Korten, David. *When Corporations Rule the World.* San Francisco: Kumarian Press and Berrett-Koehler Publishers, 1995. A book that everyone should read in order to understand what is happening to the planet Earth and the political and economic world as the corporations take possession of the planet both in its physical reality and its political controls. We are in the age when the nation-state has so declined that it must now be considered mainly as an instrument to assist the transnational corporations in their quest to control the planetary process in its every aspect.

———. *The Post-Corporate World: Life After Capitalism.* San Francisco: Kumarian Press and Berrett-Koehler, 1998. Here the author proposes a new, more organic approach to the life process, a truly democratic life based on local economies free from the oppresive dominace of the great corporations. He goes far beyond generalizations into detailed consideration of all the various aspects of how a valid economy would function. The most comprehensive guide that I have seen both in the principles presented and the details of application.

Lawlor, Robert. *Voices of the First Day: Awakening the Aboriginal Dreamtime.* Rochester, VT: Inner Traditions, 1991. An excellent collection of information and understanding of the inner world of indigenous peoples as expressed in the language and in the artistic imagery of the Aborigines of Australia.

Lebon, J. H. G. *An Introduction to Human Geography.* New York: Capricorn Books, 1966. Studies in geography are among the best ways of entering the field of ecology, as witness recent analyses of economic and political geography. Of books dealing with human geography this volume, though written some years ago, remains one of the most helpful.

Kimbrell, Andrew. *The Human Body Shop: The Cloning, Engineering, and Marketing of Life*. Upland, Pa.: Diane Publishing, 2nd ed. 1999. Orig. ed. 1993. A thorough critique of the ever-increasing atttack on the human person through reduction of the human body to a commodity to be controlled, dissected, bought, and sold as any other physical being. Of special concern are the efforts being made to intervene in the genitic process with totally inadequate understanding of how it functions or the consequences of such actions. The author, founder and director of the International Center for Technological Assessment, is remarkably clear in his thinking and precise in his writing.

Legge, James, tran. *Sacred Books of China, Part III, Sacred Books of the East,* vol. 27. Delhi: Motilal Banarsidass. Reprint 1966. (originally published in London in 1885.) This is a translation of the ancient Book of Ritual, known as the *Li Chi* in China.

Leiss, William. *The Domination of Nature.* New York: George Braziller, 1972. A study of scientific discoveries inspired by the quest for human domination of the natural world and its consequences on both the natural world and human society itself.

Leopold, Aldo. *A Sand County Almanac, With Essays on Conservation From Round River.* New York: Oxford University Press, 1966. A classic treatise, especially renowned for its section on "A Land Ethic."

Little, Charles E. *The Dying of the Trees.* New York: Viking Press, 1995. A soul-shaking survey of the condition of the forests throughout the North American continent by an author fully competent to write on this subject. Simply recalling the American chestnut and the American elm of the past and the present declining situation of the hemlock, the dogwood, the beech, and the sugar maple should evoke an intense reaction toward protecting the remaining species, which could be imperiled as long as present conditions persist.

Lopez, Barry Holstun. *About This Life: Journeys on the Threshold of Memory.* New York: Alfred A. Knopf, 1998. A collection of essays indicating just how we need to reshape our imagination and memory in our experience of the natural world.

———. *Arctic Dreams: Imagination and Desire in a Northern Landscape.* New York: Bantam Books, 1988. This book is a masterful description of Lopez's experiences in the Arctic regions.

———. *Of Wolves and Men.* New York: Scribner's, 1978. The author has an amazing ability to write from a personal sensitivity toward other species who share the planet with us.

Lovins, Amory. *Soft Energy Paths: Toward a Durable Peace.* San Francisco: Friends of the Earth International, distributed by Ballinger, Cambridge, MA. 1977. A widely read and influential presentation of a way to fulfill our energy needs without committing ourselves to nuclear sources or polluting industries.

Macy, Joanna. *Mutual Causality in Buddhism and General Systems Theory: The Dharma of Natural Systems.* Albany, NY: State University of New York, 1991. A special resonance is experienced between these two worldviews, revealing something profound in the functioning of the natural world in both its physical-material and psychic-spiritual aspect.

Mackinder, Sir Halford John. *Democratic Ideals and Realities.* London: Constable and Co., 1909. His essay on "The Geographical Pivot of History" was given in 1904. This book on democratic ideals was an expansion of his thesis. He taught at the School of Geography at Oxford, until 1904, then at the London School of Economics.

Marsh, George Perkins. *Man and Nature: Or Physical Geography as Modified by Human Action.* Cambridge: Harvard University Press, 1965. Originally published in 1864. Marsh was among the earliest American writers to give careful attention to the deleterious impact that humans were having on the natural world.

Maser, Chris. *Global Imperative: Harmonizing Culture and Nature.* Walpole, NH: Stillpoint Publishing, 1992. A scholar, forester, and thinker, the author has written an overall view of the human presence to the natural world, with indications of the need for a greater appreciation of the natural world for its aesthetic and spiritual as well as economic value to humans.

McDaniel, Jay B. *With Roots and Wings: Christianity in an Age of Ecology and Dialogue.* Maryknoll, NY: Orbis Books, 1995. A leading Christian theologian highlights important new directions for Christianity in relation to other religious traditions and in light of the environmental crisis.

McKibben, Bill. *Hope, Human and Wild: True Stories of Living Lightly on the Earth.* Boston: Little, Brown, 1995. After dealing in other books and essays with the deleterious aspects of what is happening on the planet, the author in these three essays deals with positive achievements, especially in Curitiba, a city in Brazil, and in Kerala, a state in southwest India. These are stories of ecological success in human communities.

McLuhan, T. C. *Touch the Earth: A Self-Portrait of Indian Existence.* New York: Promontory Press, 1971. A collection of remarkable statements of outstanding Indian personalities. These statements reveal the depth of insight and feeling of the indigenous peoples of the North American continent during the period of European occupation.

McPhee, John. *Annals of the Former World.* New York: Farrar, Straus & Giroux, 1998. An extremely valuable resource for its description of the geological record of the North American continent by someone with a lifelong dedication to understanding this continent in its historical emergence, its most basic structure and its integral functioning.

Meadows, Donella H., et al., eds. *The Limits to Growth: A Report for the Club of Rome's Project on the Predicament of Mankind.* New York: Universe Books, 1972. This work, sponsored by the Club of Rome and the Massachusetts Institute of Technology is, it seems, the first carefully studied overview of the human economic situation in relation to the resources of the natural world. That there was a problem remained largely unrecognized until after World War II. This book evoked intense opposition, yet the validity of its basic conclusions has been sustained. With Rachel Carson's *Silent Spring* in 1962 and the meeting of the United Nations Conference on the Environment held in Stockholm in 1972, this book can be considered as one of the main inspirations of the environmental movement throughout the world.

Meadows, Donella H., et al., eds. *Beyond the Limits: Confronting Global Collapse, Envisioning a Sustainable Future.* Post Mills, VT: Chelsea Green Publishing Company, 1992. A successor volume to the above title, this review of the subject some twenty years later reaffirms the basic conclusions arrived at earlier, with a greater sense of urgency toward reshaping the human venture in a more benign relation with the natural world.

Meiss, Millard. *Painting in Florence and Sienna after the Black Death.* Princeton, NJ: Princeton University Press, 1951. A survey of the changes in artistic expression and in forms of spirituality after the Black Death, based on extensive study of the contemporary documents of the period. An invaluable source for understanding the deeper cultural consequences of this experience, which has influenced Western history into our own times.

Merchant, Carolyn. *The Death of Nature: Women, Ecology, and the Scientific Revolution.* San Francisco: Harper San Francisco, 1990. These two books (see below) present the thought of one of the most effective historians of environmental thought to emerge from the feminist movement. The author emphasizes both social concerns and the integrity of the human community with the ecosystems of the planet.

———. *Ecological Revolutions: Nature, Gender, and Science in New England.* Chapel Hill, NC: University of North Carolina Press, 1989. A detailed study of the sequence of transformations in America in relation to the land.

Middleton, Susan, and David Littsschwager. *Witness: Endangered Species of North America.* Introduction by E. O. Wilson. San Francisco: Chronicle Books, 1994. A large-format book of photographs of many of the endangered species of North America, providing insight into what we are losing as various living forms become extinct.

Milbrath, Lester W. *Envisioning a Sustainable Society: Learning Our Way Out.* Albany, NY: State University of New York Press, 1989. A political scientist presents in three sections the results of his lifetime of thought concerning the present human situation: the present impasse, the future as an integral sustainable presence of humans in the natural world, and the way to achieve the desired transformation. The emphasis is on the need to change human ways of thinking so as to create a sustainable future.

Miller, Perry. *Errand into the Wilderness.* Cambridge: Harvard University Belknap Press, 1993. This author is the foremost scholar of the early years of New England Puritan theology and cultural thinking. Superb insight into the problems associated with the coming of European religion and culture into the wilderness world of America. He sees religion as the basic issue of civilization in relation to nature.

Montessori, Maria. *To Educate the Human Potential.* Oxford, England: Clio Press, 1948. The first woman to obtain a medical degree in modern Italy,

Montessori took on the issues associated with the integral development of children from their earliest years. She is among the most distinguished educators of our times. This might be considered one of her most significant books, especially for its insight into the importance of the child awakening to the surrounding universe.

Muir, John. *The Wilderness World of John Muir.* Edited with an introduction and interpretative comments by Edwin Way Teale. Illustrated by Henry B. Kane. Boston: Houghton Mifflin, 1954. The editor, a New England naturalist, has selected the most significant writings of the person who first appreciated in real depth the wilderness regions of northern California.

Nash, Roderick. *Wilderness and the American Mind.* New Haven: Yale University Press, 1967. 3d ed., 1989. Hardly any subject is more in need of elucidation than the meeting of persons from the Christian-humanist culture of Europe with the indigenous peoples and culture of the North American continent. The author reveals the difficulty Europeans encountered in their efforts to respond to the sacred in one of its most dramatic self-presentations on the American continent.

Neihardt, John G. *Black Elk Speaks: Being the Life Story of a Holy Man of the Oglala Sioux.* New York: Simon and Schuster, 1932. One of the most complete and authentic accounts that we have of a religious personality of the indigenous peoples of this continent. John Neihardt had the literary skill to enable the story of Black Elk to come through in the English language with exceptional clarity.

Noble, David F. *America by Design: Science, Technology, and the Rise of Corporate Capitalism.* New York: Alfred A. Knopf, 1977. This is one of the most significant works dealing with the rise of scientific technologies and the modern corporation in the late nineteenth century. It is most helpful in understanding the transition into the modern industrial period.

Noble, David W. *The Eternal Adam and the New World Garden: The Central Myth in the American Novel Since 1830.* New York: Braziller, 1968. The author describes the myth of the millennium to be realized in the newly discovered North American continent is presented as the central focus of American culture. This myth finds a diversity of expression in the various writers who have dominated the literary scene in America through the years.

Norberg-Hodge, Helena. *Ancient Futures: Learning from Ladakh.* San Francisco: Sierra Club Books, 1992. A remarkable work by a remarkable person, who is one of the few accomplished Western scholars of the Ladakh language. Because the Ladakh culture was relatively uninfluenced by modern Western culture until recently, the author has experienced the sustainable culture of this independent people in the severe climate of their original life situation. She suggests ways to maintain the culture and points toward ways that industrialized societies might learn from Ladakh.

Novak, Michael. *Toward a Theology of the Corporation: American Enterprise Institute, 1990.* New York: Penguin Books, 1986. This author presents a view of utmost praise for the modern corporation. He endows the modern corporation with a kind of sacred aura.

Ohmae Kenichi. *The End of the Nation-State: The Rise of Regional Economics.* New York: The Free Press, 1995. A clearly written description of the transition from nation-based economies to the new global economy with its consequent disruption of the earlier nation-based economies. An excellent presentation of the dissolution of the nation-state as the primary ordering principle in the public life of Western society.

Orr, David F. *Ecological Literacy: Education and the Transition to a Postmodern World.* Albany, NY: State University of New York Press, 1992. Orr's basic thesis is that education is central to creating a viable future for both the human community and for the natural life-systems of the Earth. He offers guidance in designing an educational program that will prepare students for their role in shaping a truly integral Earth community.

Osborn, Fairfield. *Our Plundered Planet.* Boston: Little, Brown, 1948. One of the first surveys of the planetary devastation taking place through the plundering process of the industrial world. In 1953 Osborn wrote another work, *Limits of Our Earth.*

Peck, Robert M. *Land of the Eagle: A Natural History of North America.* New York: Summit Books, 1990. The quotations from Captain John Smith are taken from *The Generall Historie of Virginia, New England and the Summer Isles,* fac. ed. Glasgow: University of Glasgow Press; London: Macmillan and Co., 1907, pp. 44–47. Robert Peck's book is a reliable general natural history of the continent, its geological formation, its wildlife, and its European settlement.

Ponting, Clive. *A Green History of the World: The Environment and the Collapse of Great Civilizations.* New York: Penguin Books, 1991. A comprehensive series of studies on the course of human affairs in relation to the natural world from earliest times to the present. There is both a fullness and a conciseness in the data, a clarity in the presentation, and a comprehensive range in the thinking that makes this book fascinating to read.

Prucha, Francis Paul, ed. *Americanizing the American Indian: Writings by the "Friends of the Indians" 1880–1900.* Cambridge: Harvard University Press, 1973. A valuable record of the awkward, humiliating, and often cruel efforts made toward the end of the nineteenth century to absorb the indigenous peoples of this continent into the intellectual, cultural, and religious orientation of European settlers.

Redfern, Ron. *The Making of a Continent.* New York: Times Books, 1983. A detailed, geological-based presentation of the shaping of the North American continent, in a large-book format, with photographs. The various parts of the continent are related to the formation of the North American tectonic plate, one of the complex of Continental plates that took shape with the breakup of the original Pangaea some 200 million years ago.

Register, Richard. *Ecocity Berkeley: Building Cities for a Healthy Future.* Berkeley: North Atlantic Books, 1987. A study of the city of Berkeley and how it might be periodically redesigned in its various areas over a period of a hundred years so that the original streams would be brought above ground, woodlands would be replanted, and wildlife reintroduced. People would live closer to their work. They would mostly walk or bicycle wherever they needed to go. The automobile would be extensively eliminated. Waste would largely be purified locally. With all these alterations the population could be maintained at its present level.

Reisner, Marc. *Cadillac Desert: The American West and Its Disappearing Water.* New York: Penguin Books, 1986. Study (also made into a film) of government-financed projects for damming the great rivers of the West. The story is told with all of its intrigue and with an understanding of both the natural life-systems of the West and the human communities that support these engineering projects.

Rolston, Holmes. *Environmental Ethics: Duties to and Values in the Natural World.* Philadelphia: Temple University Press, 1988. A comprehensive treatment of the subject of environmental ethics with an emphasis on the intrinsic value of nature.

Roszak, Theodore. *The Voice of the Earth: An Explanation of Ecopsychology.*

New York: Simon and Schuster, 1992. An astute commentator on contemporary affairs, Roszak defines an emerging field of ecopsychology with a profound insight into the integral relatedness of the human with the natural world.

Ruether, Rosemary Radford. *Gaia and God: An Ecofeminist Theology of Earth Healing.* San Francisco: Harper San Francisco, 1992. One of the leading feminist theologians details the potential resources in the Jewish-Christian traditions for an ecofeminist understanding of the planet and its human presence.

Ryley, Nancy. *The Forsaken Garden: Four Conversations on the Deep Meaning of Environmental Illness.* Wheaton, Illinois: Quest Books, 1998. After enduring environmental illness for many years, Nancy Ryley has written an account of her experience and her quest for understanding of the deeper sources of it in the disruption we are causing in the integral functioning of the Earth. This book is an account of the conversations she had with Laurens van der Post, Marion Woodman, Ross Woodman, and Thomas Berry on this subject.

Sale, Kirkpatrick. *Dwellers in the Land: The Bioregional Vision.* San Francisco: Sierra Club Books, 1985. This author is deeply committed to a future dependent on the intimacy of the local community with the surrounding natural setting. He sees the destructive consequences inherent in the industrial way of life.

Schmidheny, Stephan. *Changing Course.* Cambridge: MIT Press, 1992. The author has been one of the leading figures in establishing corporation control of the global economy through the World Business Council for Sustainable Development. This council, in alliance with the World Trade Organization and other world organizations, presents itself as interested in securing an integral mode of human presence to the Earth by "changing course," as it were. The difficulty is in the inherent exploitative nature of the corporations' activities.

Schumacher, E. F. *Small Is Beautiful: Economics as if People Mattered.* New York: HarperCollins, 1989. Originally published, 1973. A gracious yet powerful presentation of the need to observe a sense of scale in every phase of human life but especially in economics. Schumacher had extensive experience in planned national economies while he was in charge of the British Coal Board economy after World War II. He saw the danger of globalization that would neglect the local community on which all things human needed to be based.

Simon, Julian L. *The Ultimate Resource.* Princeton, NJ: Princeton University Press, 1981. Julian Simon is the most radical of those who see environmentalists as enemies of all reasonable thinking. He sees no population problem, no pollution problem, no resource problem that is not being fully resolved in the ordinary processes of contemporary science and economics. He feels all aspects of life and environment are constantly improving and that environmentalists are simply doomsayers in the midst of ever-increasing abundance.

Smart, Bruce, ed. *Beyond Compliance: A New Industry View of the Environment.* Washington, D.C.: World Resources Institute, 1992. An effort by the industrial world to defend itself against the accusation that industry is indifferent to the pollution of the environment that it is causing. Supplies an extensive listing of the larger and more polluting industries and the efforts they are making to diminish the deleterious consequences of their activities.

Smith, Page. *The Rise of Industrial America: A People's History of the Post-Reconstruction Era.* Vol. 6. New York: Penguin Books, 1984. A fascinating description by a master historian of the transition of America from the agricultural society of the pre–Civil War years to the industrial-technological urban society under control of modern corporations that developed in the three last decades of the nineteenth century.

Snyder, Gary. *The Practice of the Wild: Essays.* San Francisco: North Point Press, 1990. Some of the best essays by one of America's finest nature writers and poets. Snyder has established himself as someone with insight such as few attain into wildness as the deep reality of all authentic existence.

Spengler, Oswald. *The Decline of the West.* 2 vols. New York: Knopf, 1926. This book, written by Spengler before the beginning of the World War I but revised and published in the years after the war, collapsed much of the optimism concerning the future course of human affairs generated by the Enlightenment Age of seventeenth through nineteenth centuries.

Spretnak, Charlene. *The Resurgence of the Real.* Reading, MA: Addison Wesley, 1997. A critique of the modern world with its reduction of living processes to mechanism, its suppression of the local and the intimate in favor of the general and the standardized, its sanitized and sterilizing medicine that neglects the self-strengthening and healing forces of the body. The spontaneous creative powers of nature are being rediscovered. The author has presented all this with a rare clarity and grace of expression in her writing.

Spurr, Stephen H. *American Forest Policy in Development.* Seattle: University of Washington Press, 1976. A rather prosaic, official account of the attitude of the government agency toward the forests of the country. As might be expected, the crass attitude manifested toward the forest is based on the sense of the natural world as simply so much resource, so much commodity to be managed, bought, and sold like any other commercial item.

Steingraber, Sandra. *Living Downstream: An Ecologist Looks at Cancer and the Environment.* Reading, MA: Addison Wesley, A Merloyd Lawrence Book, 1997. A remarkable personal story of investigation into the chemical pollution of the environment by a scientist. Presents overwhelming evidence for the cancer-causing chemical pollution of the air, water, and soil.

Stone, Christopher D. *Should Trees Have Standing?* Los Altos, CA: W. Kaufmann, 1974. A brief study arguing for the rights of natural modes of being. Stone was supported in his position by Associate Justice of the Supreme Court William O. Douglas, who has also written on the subject.

Strickland, William. *Journal of a Tour of the United States of America 1794–1795.* Edited by Rev. J. E. Strickland, 1971. The New York Historical Society. Library of Congress Catalog Card Number: 75–165767. Precise observations concerning the settlers of the upper Hudson River Valley and their treatment of the land in the late eighteenth century.

Swimme, Brian. *The Hidden Heart of the Cosmos. Humanity and the New Story.* Maryknoll, NY: Orbis Press, 1996. In the depth and comprehensive range of his insight, in the literary quality of his writing, in the precision of his language, the author is unsurpassed in his presentation of the discoveries of contemporary science concerning the origin, structure, and functioning of the universe. He suggests ways in which we can relate to that all-nourishing abyss from which the universe is constantly coming into being.

———. *The Universe Is a Green Dragon.* Santa Fe, NM: Bear and Co., 1985. A fascinating series of discussions on how we understand the universe, and our place and function in it.

Swimme, Brian, and Thomas Berry. *The Universe Story: From the Primordial Flaring Forth to the Ecozoic Era: A Celebration of the Unfolding of the Cosmos.* San Francisco: Harper San Francisco, 1994. Perhaps the first presentation of the evolutionary process narrated in story form. Based on the authentic scientific data of a scientist committed to the study of the large-scale structure of the universe in association with a historian of cultures.

Sykes, Sir Percy. *A History of Exploration: From Earliest Times to the Present Day.* London: Routledge and Kegan Paul, 1949. 3d ed. A comprehensive survey of the manner in which the different peoples of Earth, after dividing from each other from earliest times and shaping themselves in diverse cultural modes of expression, have been discovering one another throughout historical time.

Tarbell, Ida M. *The History of the Standard Oil Company.* Revised edition edited by David M. Chalmers. New York: W. W. Norton, 1969. Originally published in 1904. The story of the corporation that, in some sense, began the modern economic-industrial world with all its ruthless power and effort to crush any competition. Originally published by *McClure's*, the most prominent journal of the period, between 1902 and 1904.

Tattersall, Ian. *Becoming Human: Evolution and Human Uniqueness.* New York: Harcourt Brace, 1998. The human story told from the Cro-Magnon period, with special emphasis on our capacity for thought, language, and symbolic activities. Of particular interest are the sections dealing with the distinctive nature of humans in relation to other animal forms.

The Tarrytown Letter. Tarrytown, NY: Published by the Tarrytown Group since 1981. Self-described as a "Forum of New Ideas," it seeks to assist business executives in carrying out their managerial functions in the ever-changing circumstances of the contemporary world.

Taylor, Bron Raymond, ed. *Ecological Resistance Movements.* Albany, NY: State University of New York Press, 1995. A study of ecological-based resistance movements throughout America, Europe, Asia, Africa, and South America.

Teilhard de Chardin, Pierre. *The Phenomenon of Man.* New York: Harper and Row, 1959. One of the earliest and most significant studies of the evolution of the universe and its culmination in the spiritual world of the human. Of special significance for its religious-theological implications. A new translation has been completed by Sarah Appleton-Weber. Brighton, Sussex, UK: Sussex Academic Press, 1999.

Templeton, John. *Is Progress Speeding Up?: Our Multiplying Multitudes of Blessings.* Philadelphia: Templeton Foundation Press, 1997. An extreme presentation of the benefits of the modern industrial world and its supposed advantages, with apparently no concern for the devastation of the natural world that is taking place and the limited prospects for the long-range future. Author is apparently a devoted disciple of Julian Simon's glowing vision of the future that will emerge out of our present scientific, technological, industrial way of life.

Toqueville, Alexis de. *Democracy in America.* Translation by George Lawrence. New York: Doubleday Anchor Books, 1969. (Original edition 1835–1840.) Considered by some competent persons as possibly the best book ever written as an interpretation of the American venture and the formation of American culture.

Toulmin, Stephen. *The Return to Cosmology: Postmodern Science and the Theology of Nature.* Berkeley: University of California Press, 1982. A survey of the status of the field of cosmology toward the end of the twentieth century based on a series of brief studies of individual cosmological thinkers in that period.

Tucker, Mary Evelyn, and John Berthrong, eds. *Confucianism and Ecology: The Interrelation of Heaven, Earth and Humans.* Cambridge: Center for the Study of World Religions and Harvard University Press, 1998. This is the first collection of essays to be published on this topic. It documents the remarkable sensitivities toward the natural world evident in Confucian thought through means of self-cultivation, ritual practice, social orientation, and political theory.

Tucker, Mary Evelyn, and John Grim, eds. *Worldviews and Ecology.* Maryknoll, NY: Orbis Books, 1994. Different worldviews have shaped the various civilizations and are still powerful determining forces throughout the human community. This collection of studies brings these traditions more directly into an awareness that each arose in the beginning by its intimate relation with the natural world and that each has a role to play in the future of the human project.

Tucker, Mary Evelyn, and Duncan Williams, eds. *Buddhism and Ecology: The Interconnection of Dharma and Deeds.* Cambridge: Center for the Study of World Religions and Harvard University Press, 1997. A major contribution to our understanding of the role of Buddhist thought in understanding the interrelated nature of the universe. The essays draw on both Buddhist theory and practice to suggest ways to live in greater harmony with nature.

Turnbull, Colin M. *The Forest People.* New York: Simon and Schuster, 1961. An account of one of the pygmy people of West Africa, in their relation to the forest. Their intimacy with the forest as a personal numinous presence has a power and unique quality that is found with indigenous peoples but tends to be lost in urban settings.

Turner, Frederick, J. *Beyond Geography: The Western Spirit Against the Wilderness.* New Brunswick, NJ: Rutgers University Press, 1992. An extensive presentation of the deep suspicion toward the natural world as seducer of human spiritual qualities throughout the course of Western civilization. Although the thesis proposed is widely valid and deserves presentation, the author fails to indicate the intimate appreciation of the natural world manifested up through the medieval period. This appreciation we find especially in persons such as Hildegard of Bingen in the twelfth century, also by Saint Bonaventure in the thirteenth century.

Van den Bergh, Jeroen C.J.M., et al., eds. *Toward Sustainable Development: Concepts, Methods, and Policy.* Covelo, CA: Island Press, 1994. An effort to bring the ecological study of economics into scientific form. Among the concepts offered is that of "natural modes of being" as "capital" in the economic sense. Also the phrase "sustainable development" as a term comes under intensive scrutiny. These papers were delivered at a workshop held at Stockholm University in 1992 under the auspices of the International Society for Ecological Economics.

van der Post, Sir Laurens. *A Far-Off Place.* New York: Harcourt Brace, 1974. A writer with a depth of insight into twentieth-century cultural developments in Africa and the European-American worlds. As a child he grew up in intimate contact with the Bushmen of South Africa. This background enables him to write with a special feeling for our human presence in the natural world and provides special insight into the world of the Bushmen as they were passing from their earlier tribal culture into the European-American civilization.

Weatherford, Jack. *Indian Givers: How the Indians of the Americas Transformed the World.* New York: Ballantine, 1988. A significant reorientation of our view of the place of the indigenous peoples of the Americas in the larger course of human history. The original inhabitants here have been immensely more significant in the history of the modern world than we have ever understood or appreciated. Outlined here are the contributions to economic development of the European empires from this continent's gold and silver; to medicine, through quinine and cocaine; to the food supply of the world; and to spiritual traditions.

Weisman, Alan. *Gaviotas: A Village to Reinvent the World.* White River Junction, VT: Chelsea Green Publishing, 1998. A detailed narrative of a village situated on the arid plains of Colombia, South America, that with a visionary leader competent in contemporary technologies and ancient cultivation gave new life to itself, and even set up conditions that enabled the tropical forest to reestablish itself.

Wilshire, Bruce. *The Moral Collapse of the University.* Albany, NY: State University of New York Press, 1990. A superb explanation of what has happened to the contemporary university and why it no longer communicates to its students a world of meaning. This loss of larger context of meaning in our highest levels of intellectual development leaves our society destitute of those real values that human life should have.

Wilson, E. O. *Biophilia.* Cambridge: Harvard University Press, 1984. An absorbing description of the manner in which living beings are attracted to each other, told by a biologist of unique gifts, as a scientist and as a writer with something akin to poetic expression in some of his essays.

———. *Consilience: The Unity of Knowledge.* New York: Alfred A. Knopf, 1998. A distinguished biologist and strong proponent of empiricism, Wilson expresses his thoughts on the unity of knowledge. Although it has all the inherent inadequacies of empiricism, this work belongs among the more significant writings of modern scientists in their effort to explain themselves and their work in the larger perspective of human understanding.

Wilson, E. O., ed. *Biodiversity.* Washington, D.C.: National Academy Press, 1988. A collection of over fifty essays by extremely competent scientists concerning the manner in which the life-systems of the planet function, indicating the damage already done and the future consequences of ignoring the basic pattern of interdependent life-forms. These essays are evidence enough that we know how humans should relate to the life-systems of the planet, even though we will always be limited in our insight.

World Commission on Environment and Development. *Our Common Future.* New York: Oxford University Press, 1987. The commission, under chairperson Gro Harlem Brundtland, was authorized by the United Nations to inquire into the problems associated with the deleterious consequences of the industrial process as it is extended throughout the human community. This is the report. It was followed by the United Nations Conference on Environment and Development held in Rio de Janeiro in 1992.

World Conservation Union. *Caring for the Earth: A Strategy for Sustainable Living.* Gland, Switzerland: World Wildlife Fund, 1991. A study sponsored by the United Nations Environmental Program, the World Conservation Union, and the World Wide Fund for Nature, it articulates the principles and conditions for sustainable living, the special needs for various human situations, with indications of the manner in which the strategy can be carried out. In the precision of its proposals, and the design for their fulfillment by the community of nations, this document is one of the most significant to come from these official sources.

Worster, Donald. *Nature's Economy: A History of Ecological Ideas.* 2d ed. Cambridge: Cambridge University Press, 1994. The basic work dealing with the origins and development of the study of ecology. Our understanding of the natural world and the human role within the natural world had a new beginning with the scientific discoveries of the seventeenth century and our progressive understanding of the universe and the planet Earth as an evolutionary process.

———. *Rivers of Empire: Water, Aridity, and the Growth of the American West.* New York: Pantheon Books, 1985. A study of the development of the western regions of the North American continent by a people indifferent to the ecological structure and functioning of the region, which had the result of terrible consequences to the land and its ecosystems.

Wub-e-ke-niew. *We Have the Right to Exist: A Translation of Aboriginal Indigenous Thought. The First Book Ever Published from an Ahnishinahbæó'jibway Perspective.* New York: Black Thistle Press, 1995. One of the most authentic and complete responses from the indigenous peoples of North America to the intrusion and devastation of their culture by the incoming peoples and cultures of the European world. The Euro-Americans still cannot quite appreciate just how unjust and oppressive their occupation of the continent has been and remains, even in the present.

Yergin, Daniel. *The Prize: The Epic Quest for Oil, Money, and Power.* New York: Simon and Schuster, 1991. A monumental narrative of over 800 pages of the beginning and development of the petroleum industry and its influence on both the economic and political spheres of the human community.

Index

44² 56¹

About the Author

THOMAS BERRY, a historian of cultures, comes from the hill country of the Southern Appalachians, where he was born at the beginning of the First World War. In 1934 he entered a monastery. He received his doctoral degree in Western cultural history at the Catholic University of America in Washington, D.C., in 1948. In that year he went to China to study Chinese language, culture, and religions. After Mao Tse-tung came to power, he returned to the United States to continue his Chinese studies and begin his study of the Sanskrit language and the religious traditions of India. He has taught at the Center for Asian Studies at Seton Hall University and at the Center for Asian Studies at Saint John's University in New York.

He was the director of the graduate program in the history of religions at Fordham University from 1966 until 1979. Founder of the Riverdale Center of Religious Research in Riverdale, New York, he was its director from 1970 until 1995. During this time the Center's major concern was to clarify the role of the human community within the more integral community of the Earth and with the universe itself. He was president of the American Teilhard de Chardin Association from 1975–1987. In 1995 he returned to North Carolina, where he is continuing his writing on ecological issues.

In 1968 he published *Buddhism* and in 1972 *The Religions of India.* He has also published a number of articles on the more significant human issues of the twentieth century. For the past fifteen years his writings have focused on the disturbed ecological situation of industrial societies. *The Dream of the Earth,* published in 1988, won a Lannan Award for nonfiction in 1995. *Befriending the Earth,* a series of conversations on religion and the Earth, with Thomas Clarke, was published in 1991. *The Universe Story,* with Brian Swimme, was published in 1992.